CONTROLLED REPRODUCTION
in Pigs

Controlled Reproduction in Farm Animals Series

Ian Gordon, *Emeritus Professor of Animal Husbandry, University College, Dublin, Ireland*

1: Controlled Reproduction in Cattle and Buffaloes

2: Controlled Reproduction in Sheep and Goats

3: Controlled Reproduction in Pigs

4: Controlled Reproduction in Horses, Deer and Camelids

CONTROLLED REPRODUCTION
in Pigs

IAN GORDON
BSc, MA, PhD, DipAgric (Cantab), NDA

Emeritus Professor of Animal Husbandry
University College
Dublin
Ireland

CAB INTERNATIONAL

CABI is a trading name of CAB International

CABI Head Office
Nosworthy Way
Wallingford
Oxfordshire OX10 8DE
UK

Tel: +44 (0)1491 832111
Fax: +44 (0)1491 833508
Email: cabi@cabi.org
Web site: www.cabi.org

CABI North American Office
875 Massachusetts Avenue
7th Floor
Cambridge, MA 02139
USA

Tel: +1 617 395 4056
Fax: +1 617 354 6875
Email: cabi-nao@cabi.org

A catalogue record for this book is available from the British Library, London, UK.
A catalogue record for this book is available from the Library of Congress, Washington DC, USA

ISBN-13: 978-0-85199-116-0
ISBN-10: 0-85199-116-5

Four volume set: 978-0-85199-118-4

First published 1997
Transferred to print on demand 2008

Printed and bound in the UK by CPI Antony Rowe, Eastbourne.

Contents

Preface

More than a decade has passed since the appearance of the edition of the book entitled *Controlled Breeding in Farm Animals* (Pergamon Press), on which this new four-volume series is based. The aim of this volume of the second edition is to provide a fairly detailed and up-to-date view of the literature dealing with the many different ways in which reproduction in pigs may be controlled and manipulated. The hope is that the book will prove to be of value and interest, not only to students of animal science and veterinary medicine, but also to those concerned with the practical aspects of reproduction control, whether in research, in an advisory capacity or in applying the techniques directly on the farm. Although the book does not concern itself with reproductive disorders or infertility problems, certain of the material in the text should be of interest to veterinary practitioners. For those advanced undergraduates in animal science and veterinary medicine contemplating research in reproductive physiology, the work may provide some insight into the nature and scope of current reproductive technology and of the problems that await solution.

It would be foolish to claim that any work such as this can be other than incomplete, in view of the vastness of the literature, but an attempt has been made to ensure that most statements of substance are backed by an appropriate reference. It should be emphasized that the text covers areas such as embryo transfer technology in which there is considerable research activity; for many readers, the chapters may serve as nothing more than a starting point in seeking information on their particular interests. A major objective of the work is to draw attention to that information which may be used directly to increase the worldwide efficiency of pig production systems.

The present text has been distilled from research and teaching interests in the UK, the USA and Ireland spanning a period of more than 40 years. As a graduate in Agricultural Science from Nottingham University in the early 1950s, the author had the opportunity of working as a graduate student under the late Sir John Hammond at the School of Agriculture and Animal Research Station in Cambridge. At that time, in the early 1950s, difficulties facing a researcher in reproductive physiology in Cambridge included lack of pasture and space to keep farm animals. For that reason, the author was to spend much of the 1950s out of the laboratory working directly with farmers and their sheep and cattle in many of the counties of England and Wales. Later, in Ireland, ably supported by an enthusiastic band of graduate students, the author continued to work happily with farmers in developing controlled reproduction techniques. In more recent times, the author has been confined to the laboratory, mainly in an effort to see progress in the laboratory production of cattle and sheep embryos. Nevertheless, it is with farmers and the animals they tend that the author feels most at home. It is hoped that this book will be of some value to all those who have the best interests of animal agriculture at heart.

Acknowledgements

The author wishes to express his gratitude to the many editors and publishers who have given permission to reproduce illustrations and tables from books and original articles. Sincere thanks must also be extended to the staff of CAB INTERNATIONAL for their excellent help during the preparation and production of this volume.

The author greatly appreciates the provision of office and facilities at Lyons Research Farm by University College Dublin to allow the writing to be done in the peace and quiet of the countryside.

Finally, thanks are due to a wife who cheerfully tolerated the many hours of absence involved in writing the book.

Abbreviations

ACTH	adrenocorticotrophic hormone
ADV	Aujeszhy disease virus
AI	artificial insemination
BSA	bovine serum albumin
BTS	Beltsville thawing solution
CAM	calcein acetylmethyl ester
CAP	chlormadinone
C-O-C	cumulus–oocyte complexes
CTC	chlortetracycline
DMSO	dimethyl sulphoxide
EDTA	ethylenediaminetetraacetate
EH	ethidium homodimer
EGF	epidermal growth factor
EIA	enzyme immunoassay
EPF	early pregnancy factor
ESR	oestrogen receptor marker
ET	embryo transfer
EU	European Union
FCS	fetal calf serum
FSH	follicle stimulating hormone
GIFT	gamete intrafallopian transfer
GnRH	gonadotrophin releasing hormone
GV	germinal vesicle
hCG	human chorionic gonadotrophin
ICM	inner cell mass
IGF-I	insulin-like growth factor
IU	international units
IVF	*in vitro* fertilization
IVM	*in vitro* maturation
LH	luteinizing hormone
MAP	medroxyprogesterone acetate

MLC	Meat and Livestock Commission
MMA	metritis, mastitis and agalactia
PBS	phosphate-buffered saline
PEF	pig fetal endometrial fibroblasts
$PGF_{2\alpha}$	prostaglandin $F_{2\alpha}$
PMSG	pregnant mare serum gonadotrophin
POEC	pig oviductal epithelial cells
PPV	porcine parovirus
PVA	polyvinylalcohol
RIA	radioimmunoassay
SFV	swine fever virus
SPA	sperm penetration assay
SPF	specific pathogen free
TNF	tumour necrosis factor
TRH	thyroid stimulating hormone
USDA	United States Department of Agriculture
UTJ	uterotubal junction

Introduction to Controlled Breeding in Pigs

1

1.1. Introduction

The efficient production of lean meat for the consumer is the primary objective of pig production and the economics of pig production depend very much on reproductive efficiency (English *et al.*, 1984). Pig production in Ireland and the UK, as in many other countries, is currently based primarily on intensive systems, designed to make the maximum use of space within controlled environment buildings with minimum labour requirements. Such production systems have proved very effective in creating an efficient pig industry, with the ability to supply consumers with a high quality product at low cost. In contrast to cattle and sheep, which have a relatively low reproductive output annually, the pig has the ability to produce a large number of young in a short space of time; certainly, given appropriate management, housing and nutrition, pigs can prove to be among the most profitable of the meat-producing animals.

The usual method of assessing reproductive performance, which can vary markedly from farm to farm, is to examine the number of young produced each year by the sow. This characteristic comprises two important components: litter size and the number of litters produced each year. The fact that there can be considerable variation from herd to herd suggests that many factors are involved and that some of these are under the control of the farmer. It is clearly important to understand the effect of factors such as breeding, feeding and management on reproductive performance under normal commercial conditions. In addition, there is the need to examine carefully any technique which may be capable of increasing the reproductive efficiency of the pig herd.

Pigs: ancient and modern

In the beginning of its history, which probably goes back some 7000 years, the pig was a forest-dwelling animal. It is believed that the European breeds of domestic pigs were derived from the local wild pig, *Sus scrofa*; herds ranged over pastures and through forests and were only kept indoors for fattening. The breeds in the Far East were apparently derived from another wild pig, *Sus vittatus*, a smaller animal with shorter legs and a higher reproductive ability (Signoret *et al.*, 1975); the two types will readily interbreed. The modern breeds of pig are believed to have evolved from different crossings between these two original types. The pig is a member of the Artiodactyla, or even-toed ungulates; it has four toes on each foot, but only the third and fourth are functional. The feature that all pigs possess is a flat, rounded snout on a flexible muzzle. Another feature common to all pig breeds is the modification of the upper canine teeth into tusks. Although the hair is sparse, the body of the pig is well insulated by a layer of subcutaneous fat; the skin temperature tends to follow that of its immediate surroundings. The pig's characteristic body odour is produced by skin gland secretions; the carpal glands are well differentiated on the front legs. In the boar, secretions of the preputial pouch give the male its characteristic 'boar' odour. The pig has a keen sense of smell which it uses in rooting for buried food; advantage is taken of this in the Perigord region of France in training pigs to search for truffles (subterranean mushrooms).

In many parts of the world, there was a marked parallelism in the populations of pigs and poultry; both animals tended to draw their sustenance from identical sources, i.e. from the scavenging of the village and the countryside. As a consequence of its omnivorous habit and ready adaptability, the pig is able to survive well under a wide range of conditions. Both as a free-range animal and one kept under close confinement, it has long been a useful converter of waste into meat. The distribution of the pig, however, has been markedly influenced by the religious sentiment against the consumption of pig meat, a factor which applies especially to the Muslim countries of the Middle East, North Africa and western Asia. On the other hand, in far eastern and in parts of southern Asia, pig meat consumption is high (especially in China). According to Filatov and Rybalko (1990), pig meat production in China accounted for 87% of all meat produced in that country; the Chinese population of 340 million pigs (20 million sows and 2 million boars) at that time represented some 40% of the world pig population.

At the start of the 1990s, the world's population of 839 million pigs were mainly distributed in the important pig-producing countries in North/Central America, South America, Europe, Africa, Asia and Australasia (Blair, 1990); according to this author, in the developed countries, future trends are likely to be dictated by domestic consumption and exports, whereas in the developing countries the potential for increased pig production is likely to depend on feed availability.

The Large White pig, which originated in the UK, is known in many countries as the Yorkshire; the breed played a large part in the development of the modern domestic pig. The Landrace pig has also been widely represented in the development of the modern pig.

Crossbreeding systems

Crossbreeding unrelated lines of pig breeds cashes in on heterosis, the increased performance of the cross by comparison with the mean of the parental lines. An F_1 piglet shows individual heterosis in its ability to survive the hazards of prenatal, perinatal and postnatal life. An F_1 sow shows maternal heterosis in terms of a shorter rebreeding interval, higher litter size and increased milking ability (Johnson *et al.*, 1984). Crossbreeding began in the USA fifty and more years ago and two-, three- or four-breed rotation crossing systems were adopted by the majority of producers in the pig industry in that country. As noted by Bichard and David (1986), Europe came to crossbreeding in pigs at a later stage, but had the advantage of more fertile breeds to work with and soon came to appreciate the value of the F_1 sow. In 1970, for example, every sow and every slaughter pig in Denmark was a purebred Landrace; by the mid-1980s, almost all commercial herds in that country contained F_1 (Large White × Landrace) sows, with backcross or three-breed cross litters, thereby achieving the benefits of heterosis in a very short time.

In the UK, where most of the commercial pigs are now crossbred, farmers had traditionally crossed the Large White and Landrace for the beneficial effects of hybrid vigour; in the 1970s, pig-breeding companies improved on this technique by selecting high performance lines from various pure breeds and combining these to produce first or second cross gilts for selling on to commercial producers; boar lines were also selected to mate with the gilts. In the mid-1980s, Bichard and David (1986) mentioned statistics produced by the Meat and Livestock Commission (MLC) at that time, involving more than 700 producers and 300,000 litters per year, which showed sows weaning 9.2 piglets per litter and 20.8 piglets per sow per year.

According to Webb (1995), the key objective of the last 30 years in pig improvement programmes has been to reduce backfat, but now that this has been largely achieved, breeding aims are changing; there is now pressure to improve the quality and uniformity of pig meat and to provide animals that are easier to manage in large enterprises. In seeking further progress, recent advances in molecular biology are likely to widen the options markedly. It is now possible, for example, to eliminate the porcine stress syndrome gene completely by DNA testing.

Of the various farm mammals, the pig has several features which makes it particularly well-suited to gene mapping; the short generation interval (about a year) and large litter size (10–15) enable the rapid development of effective pedigrees for linkage analysis. The pig genome is assumed to be of similar dimensions to that of the human and mouse genomes, containing up to 100,000 different genes. Cytogenetic and other physical approaches are facilitated by the heterogeneity in size and relatively small number of pig chromosomes; there are 18 pairs of autosomes together with the X/Y pair of sex chromosomes.

The European Union (EU) Pig Genome Mapping Project, together with mapping programmes operating elsewhere, has produced over 1000 potential 'marker genes' scattered over the 38 chromosomes. Although most production traits are controlled by several genes, the maps can be used to search for genes with large effects on performance; marker genes and performance genes which are close on the same chromosome tend to be inherited together. The performance genes can be selected using the neighbouring markers, in a process known as marker-assisted selection, thus leading to faster rates of progress (Haley, 1996; Visscher *et al.*, 1996). Studies reported by Meuwissen and Van Arendonk (1992) suggested that the rate of genetic gain in a breed could be increased by 10–20% when using marker-assisted selection.

Markers have already been reported for certain performance traits, including litter size. One of these, the oestrogen receptor (ESR) marker is known to carry a litter-size advantage in Meishan crosses of about 1.5 piglets per litter (Rothschild *et al.*, 1994). According to Webb (1995), markers may possibly speed up the exploitation of the Meishan litter-size advantage, by backcrossing the gene into European dam lines in a process known as marker-assisted introgression.

Chinese breeds

Of the 100 or more indigenous pig breeds in China, there are some that excel in prolificacy; the best documented of these is the Meishan, which originated around Lake Taihu in the south-east of the country. The Chinese Meishan is extremely prolific when compared with European breeds and can produce up to 30–40% more live piglets per litter. The pig was imported into the UK so that factors responsible for its extreme prolificacy (>14 piglets per litter) could be studied. Meishan crossbreds are already in use on commercial farms in the UK and there is considerable research activity at centres such as the Roslin Institute in Scotland. Other useful Chinese pig breeds include the Jinhua, originally from central China, which is famous for its meat quality, palatability and flavour; like the Meishan, the breed has the traits of early ovarian development and puberty (Kano *et al.*, 1994). Work carried out in the USA on the performance, physiology, gene mapping and crossbreeding of Meishan, Fengjing and Minzhou pig breeds with non-Chinese breeds has been reviewed by Liu *et al.* (1994).

Miniature pigs

Several strains of miniature pigs have been developed in the USA by selection from wild pigs descended from those introduced by Spanish settlers in the sixteenth century. These pigs are small compared with the domestic farm pig, but they may still attain an adult size of 70–100 kg. Small size is an advantage when the pig is used as a laboratory animal in various lines of research. According to Mount (1968), the two principal breeds of miniature pig available in the USA at that time were the Hormel and the Pitman–Moore strains.

1.1.1. Changing structure of the pig industry

The pig industry in Ireland, in common with that in many other countries, is continually moving in the direction of fewer and larger units. Irish figures have shown a marked decrease in the number of farmers keeping pigs in the recent two decades. According to official figures, the sow herd population in the Irish

Republic in 1995 was about 152,000 sows, with a capacity to produce some 3.28 million pigs in the year; there were 22 sow units with more than 1000 sows each, carrying about 34,600 sows (24% of the total breeding population). Of the 589 breeding units in the country, 73 contained more than 500 sows, making up almost half the sow population.

Elsewhere, units with 500 sows or more, once a rarity, have become commonplace; capital investment in buildings and equipment has increased greatly. According to Lean (1988), the pig industry in the UK after the Second World War was characterized by a decrease in the number of pig farmers and an almost fourfold increase in the national herd size; pig units became specialized, using relatively small acreages of land and relatively sophisticated intensive housing.

Looking towards the future of the pig in the UK, Fowler (1980) noted that feedstuff concerns, producer groups and processors were forming business arrangements which would put the British pig industry well on the road towards the vertically integrated structures of the poultry industry. The same author observed that the tendency towards larger units was likely to increase because they offered considerable scope for highly efficient management. The disposal of effluent, which is one of the major problems with pig units, large or small, may be eased by large-scale units, since these are best able to justify the costs of installations for the anaerobic digestion of waste to methane.

Changes in the structure of the pig industry in the USA in the period 1970–1990 have taken place mainly due to technical developments in nutrition and housing (Saito and Sugiyama, 1990); in general, production systems have changed from small, family-owned farms to large-scale, intensive, industrial operations. According to these authors, the US pig industry is well on its way to becoming low-cost, highly integrated and very industrial; such trends are more obvious in the corn-belt states than in the southeastern states, which are more dependent on commercial feed than the corn-belt states. Data from the United States Department of Agriculture (USDA) for 1995 show that 4.8% of producers now produce and market 43% of all pigs in the USA. It is evident from recent structural changes in the US pig industry that pig production is becoming industrialized in much the same way as the poultry industry during the 1960s. In terms of pig-breeding companies, the structure of the USA pig industry is apparently different from that in the UK, with independent breeders having 60% of the market share for breeding stock (Kerr, 1995).

1.1.2. Larger units and reproductive performance

As pig units have become larger and more specialized, the needs of production have placed increased demands on reproductive performance; at the same time, larger-scale confinement production itself may well raise new problems which have to be solved in order to achieve and maintain optimum reproductive performance. In a review of the performance of pigs in large production complexes in the former USSR, Benkov (1989) noted that reproductive efficiency had declined markedly and that attempts to improve it had been ineffective. The same author suggested that the high housing density of pigs in these large complexes may have triggered mechanisms associated with population regulation in the wild, leading to reduced fertility.

Pig production outdoors

Large-scale commercial pig production out-of-doors was pioneered in the UK during the 1950s, but herds that practised this system remained a small minority of about 5% for many years (Edwards *et al.*, 1994). By the 1990s, however, economic circumstances and animal welfare concerns about intensive pig production had led to a rapid increase in outdoor production systems; by 1991, 16% of MLC-recorded herds were outdoors and it was predicted at that time that the figure would increase to more than 20% of the national breeding herd.

It is evident that modern outdoor pig production has been associated with major changes in the genotype of the pigs as well as in the management system. According to Edwards *et al.* (1994), the traditional 'blue' sow, a crossbred incorporating the Saddleback breed, was famed for good maternal characteristics, but it has been increasingly replaced by crosses between Duroc and hyperprolific white lines, giving a sow which is hardy but produces progeny with the low fat levels required to meet market demands. An average mortality of 20% was observed in a study by Edwards *et al.* (1994), which was higher than the figures for indoor herds in the UK at that time. Although it had been expected by some that the level of stillbirths would be lower in outdoor systems than indoor ones because the piglets are able to exercise and develop good muscle tone, this was not observed; the major cause of mortality was the crushing of piglets of low birthweight which died before suckling.

In the USA, McGlone *et al.* (1994) evaluated English-style intensive outdoor pig production in the dry western plains compared with an intensive indoor system during gestation and farrowing/lactation. Based on numbers of pigs weaned per litter and weaning weights, these workers concluded that outdoor intensive pig production was a reasonable alternative to the indoor system.

In Germany, Oldigs *et al.* (1995) compared several outdoor and indoor systems of pig production over a two-year period, using two strains of German Landraces. The authors demonstrated that housing pregnant sows in an outdoor system may have some advantages in respect of animal welfare. In outdoor sytems, the introduction of new sows led to some problems, although the incidence of piglet injuries in individual housing of sows was reduced. Farrowing results tended to be better in outdoor than in indoor systems, although physiological traits indicated that sows still experienced some stress in outdoor systems.

There may be certain aspects of reproduction in sows kept outdoors that are of biological interest. European wild boars are seasonal breeders and the sows are known to exhibit anoestrus during summer (Mauget, 1985). Although this reproductive feature has waned in domestic sows, there may be a tendency towards summer infertility under some conditions (Love *et al.*, 1993). There has been a suggestion that sows culled from outdoor pig herds for failure to breed during summer may be intrinsic seasonal breeders (Bassett *et al.*, 1994). Determination of seasonal changes in plasma luteinizing hormone (LH) and prolactin indicated that declining LH secretion during the late spring/summer months may contribute to summer infertility.

Sustainable pig production systems

Sustainable pig production systems have been defined as those that combine production and management techniques to enhance profit and improve the ecological and socioeconomic surroundings (Honeyman, 1994). In the long term, the system should maintain or enhance the environment and the quality of life for the producers and society at large. Pigs are regarded as versatile animals which have many of the characteristics which will enable them to adapt to sustainable systems.

Wild boar in the UK

Kyle (1995) describes the history of wild boar farming in the UK and some other European countries; details are also given of nutrition, management, health, welfare and legal requirements. According to Kyle (1995), in early 1995, there were probably 400 breeding sows on farms with an annual production of 2000 piglets. In production terms, sows normally conceive in December–January and have an average gestation period of 114 days; litter size averages five piglets, which are weaned at 12 weeks of age. Domestic sows crossed with wild boars regularly farrow twice a year and have larger litters than wild boars, but purebreds live longer and are said to produce superior meat compared with crossbreds.

1.1.3. Animal welfare considerations

Although it is generally recognized that there is considerable potential for increasing the efficiency of pig production, including reproductive efficiency, it is essential to keep a watchful eye on animal welfare considerations. The learning ability of the pig is considerable and the social behaviour of the animal is elaborate. For such reasons, welfare problems arise in pig production when the animals are unable to control events in their environment or when they are faced with unpredictable situations. Undoubtedly, improvements in the reproductive efficiency of pigs is likely to result from a greater awareness in the industry of the causes of reproductive failure or of decreased reproductive performance. The fact that it is not only a question of buildings and feeding levels but also one of the relationship between stockperson and the pigs should help in the adoption of acceptable, industry-wide management practices relating to the conditions under which the animals are kept.

It is certain that the pig industry will have to face continued charges by informed and uninformed public opinion from the welfare lobby. The practices that operate in pig units are likely to have to meet with the approval of the stockpersons themselves as well as by society at large. In densely populated countries, such as the UK, pig production units with their odours and effluents have become a frequent target for the attentions of the urban-dwelling environmentalists. Public concern about the conditions in which farm animals are reared in intensive production units has resulted in Codes of Practice and Welfare legislation in a number of countries; this concern must be shared by farmers and by those who tend the stock, if the more extravagant demands for regulations governing animal production are to be countered effectively by the livestock industry.

1.1.4. Social environment and reproduction

It has been recognized for many years that social environment can exert a marked influence on reproduction in the pig. There is also ample evidence suggesting that certain of the difficulties encountered in modern pig production may be the result of a suboptimal social environment. It follows from this that improvements in reproductive efficiency are likely to be achieved by utilizing and increasing the existing knowledge of the role of the social environment on the reproductive processes of the animal. In the 1970s, there were writers who drew attention to the fact that the relationship formed between the stockperson and the animals could be an important factor determining their productivity but there was a general lack of objective data to substantiate the point. In pigs, Hemsworth *et al.* (1981) were among those who provided some objective information; they concluded, from their work in the Netherlands that, where the relationship between the stockperson and the pigs was poor, the sows showed a stress response which adversely affected the reproductive performance.

Hemsworth and Barnett (1990) in experiments in Australia consistently found that pigs displaying high levels of fear of humans showed clear evidence of a chronic stress response with detrimental effects on production. The major reproductive effect of high levels of fear appears to be on conception rate and litter size rather than on sexual behaviour. The fact that a good relationship between the stockperson and the breeding animal is an important factor determining reproductive performance makes it all the more important that the animal's welfare is given every consideration. In a study of relationships between stockperson and pig, Hemsworth *et al.* (1989) concluded from their evidence that reproductive performance could be improved by employing personnel that have desirable attitudinal and behavioural profiles.

Family pen system of pig production

Although there have been many efforts to improve the welfare elements of current pig production systems, some researchers have attempted a much more radical approach. Data have been presented by Wechsler *et al.* (1991) in

Switzerland of the behaviour and reproductive performance of sows housed in Stolba family pens first developed at the Edinburgh School of Agriculture (Stolba and Wood-Gush, 1984). These pens endeavour to provide natural and non-stressful surroundings for up to four sows and their litters, which remain with their dams until the birth of the next litter. In the Stolba system, which includes nest, activity and rooting areas, and which was improved after Alex Stolba's return to Switzerland in 1982, sows are reported to produce an average of 2.47 litters per year, and postnatal mortality due to crushing has been reduced from 10% to 2.4%.

Although the prototype Stolba system gave promising results in Edinburgh, a number of aspects of the system prevented it from being fully implemented in commercial practice in Scotland; these included high piglet mortality, the introduction of new breeding females within the constraints of commercial breeding policy and the high labour requirement necessary to service the system (Edwards and Baxter, 1989).

Nurtinger pig-rearing system

Schwarting and Kleiner (1995) discussed pig production in Germany in the year 2000. They describe the Nurtinger system of pig rearing in which pigs are allowed to move freely and have access to areas where thay can be undisturbed. There is an 'entertainment area' in which pigs can play in straw; scratching posts are provided and animals have access to showers for cooling. The authors suggested that pig prices will stabilize at a low level due to low feed costs, that these low costs will not affect profitability and that pig production will be capable of generating a high income.

Pen density in rearing replacement gilts

Some workers have conducted studies to investigate the effects of pen-density requirements during the growing–finishing period on the later reproductive and maternal traits of pigs. In the USA, Kuhlers *et al.* (1985) found that low pen density (eight pigs per pen) during rearing led to larger litters at birth but to decreased 21-day litter-sizes, litter-weights and survival rates. The indications were that gilts to be kept as replacement females should be raised in higher pen densities (16 pigs per pen) to increase the number of pigs surviving to three weeks. Because all replacement gilts have to go through the stress of parturiton and lactation, the authors speculated that conditioning gilts to stress earlier may be desirable.

Confinement and oestrous behaviour

The influence of housing systems for pregnant sows on their reproductive behaviour at oestrus was studied by Pedersen and Jensen (1989) in Denmark; confined and group-housed sows at a commercial farm were observed. The stress levels in late pregnancy and at oestrus were measured, using a 'fear reaction test' and plasma cortisol levels, respectively. Two weeks after oestrus, a 'social response test' was carried out, in which each sow was exposed to an unknown gilt; oestrous behaviour was detected 0.5 days earlier in the group-housed sows than in the confined animals, mainly due to differences in the sows' reaction towards humans.

In Australia, Barnett and Hemsworth (1991) examined the effects of individual and group housing on sexual behaviour and pregnancy in pigs; they record that pigs housed during the mating–gestation period in tether stalls were more restless during mating and had a significantly lower pregnancy rate (65% vs. 86%) than group-housed animals.

Good stockmanship

The existence and influence of empathy between the stockperson and the pig is now known to be a question which deserves close attention; there is ample evidence to show that the human factor can have a marked effect on the results achieved from farm animals. According to English and Burgess (1995), it is evident that education, training and the provision of a good working environment can influence the development and maintenance of empathy in pig production systems; it is also clear that much more research is required to determine the nature and type of signals that are likely to be the most useful in the attainment and maintenance of a strong stockperson–animal bond.

On the basis of experiences to date, it is likely that the remainder of the 1990s will be marked with increased confrontation between those representing animal rights and animal welfare and pig producers over the humane treatment of the animals under their care. It lies with the pig producer and the stockperson to counter such challenges by demonstrating the positive way in which they can influence the welfare and behaviour of the animals in their charge.

1.1.5. Identifying susceptibility to stress in pigs

The porcine stress syndrome (malignant hyperthermia) has a genetic basis and can be triggered in susceptible pigs by environmental stress or by administration of the inhalation anaesthetic halothane. Typical reactions involve muscle stiffness, irregular respiration, increased temperature, respiratory acidosis and death, with resulting low intramuscular pH and inferior meat quality (pale soft exudative pork). Fortunately, it has proved possible to isolate the genes for 'halothane' susceptibility and porcine stress syndrome and to develop an appropriate DNA test for use in breeding programmes to select against this condition (Russo *et al.*, 1993; Davies, 1994).

1.2. Areas of Controlled Reproduction in Pigs

Controlled reproduction in pigs, as the term is used in the present discussion, covers a range of possibilities. In the main, these possibilities relate to reproductive performance in the sow and gilt, but it is recognized that many factors can influence the efficiency of the boar in reproduction. It should perhaps be kept in mind that reproductive efficiency can be influenced, in a way much less apparent in other farm species, by the association of the sexes, whether this be in matters such as the elicitation of the full immobilization reflex in the oestrous sow (Table 1.1) or in triggering the onset of puberty in the gilt.

When considering controlled reproduction with its several aspects (Fig. 1.1), it is as well to note that certain of the control measures could probably be integrated to provide an enhanced effect. Effective oestrus control procedures, fixed-time artificial insemination and farrowing control would allow for all-in all-out systems that would improve the efficiency of labour use, reduce

Table 1.1. Effectiveness of various stimuli in eliciting the immobilization reflex in the oestrous sow (from Signoret *et al.*, 1975).

Stimulation	Postural standing reaction (%)
Pressure on back only	48
Broadcasting boar's courting grunts	70
Odour of preputial secretion at 40°C	80
Smell and sound of boar	90
Smell, sound and sight of boar	97

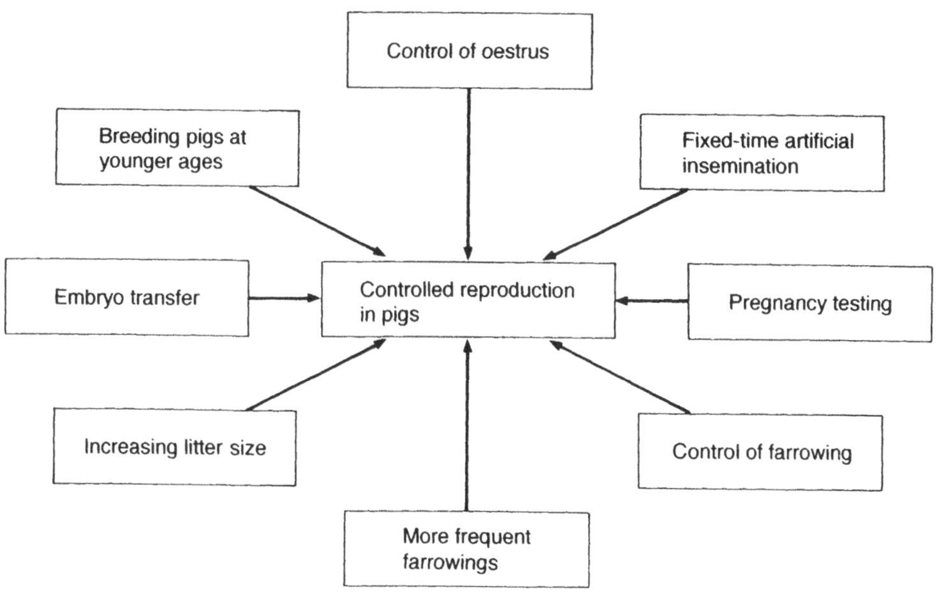

Fig. 1.1. Aspects of controlled reproduction in the pig.

the incidence of disease and the risk of stress and enable the farmer to schedule his or her entire pig operation.

In attempting to control the farrowing process in gilts within the working week and more especially, within the working day, it would be an obvious advantage to have control both of the day of mating as well as that of parturition. If effective controlled farrowing in a herd requires the administration of prostaglandin on day 114 of gestation, then this could be made that much easier if, in the first instance, the actual date of mating is controlled with that in mind. Thus, the full advantage of a technique for precisely controlling time of parturition in gilts may only be obtained when an effective method for controlling oestrus and ovulation is used. Martinat-Botte *et al.* (1985) referred to a reproductive management system being developed in France involving a sequence of events associated with synchronization of oestrus (using altrenogest), pregnancy diagnosis (with ultrasound scanning) and synchronization of parturition with prostaglandins.

1.2.1. Developments in pig breeding

Looking towards some future developments in pig-breeding programmes, it is clear that research using information from the human genome project may be valuable in pigs. The use of human probes to locate homologous pig genes is already under way in a number of laboratories. Schook *et al.* (1994) stress the need to integrate molecular genetics into the identification of major genes affecting reproduction and other important production traits in pigs. However, the pig genome contains up to 100,000 different genes and identifying those which contribute to variation in performance traits is not an easy task.

As mentioned earlier, genome mapping projects have produced more than 1000 potential 'marker genes' in the chromosomes of the pig (Webb, 1995). These markers can be used to search for genes with large effects on performance. At the Roslin Institute in Edinburgh, for example, genes affecting certain quantitative traits in pigs have been mapped; these include genes controlling growth rate and fatness in pigs.

There is also likely to be considerable interest in certain techniques associated with embryo transfer in pigs, such as cloning, sexing and the production of transgenic animals. Recent advances in cloning as applied to sheep (Campbell *et al.*, 1996) have demonstrated the possibility of cloning identical sheep by nuclear transfer from cultured cell lines. According to Solter (1996), although cloning mammals from adult cells will be considerably more difficult to achieve, it can no longer be considered impossible. On the pig front, Rothschild (1990) suggested that cloning from adult cells could prove a boon to commercial pig breeding companies who would wish to market numerous copies of superior animals.

There may well be much to be gained from the production of transgenic pigs for use in human medicine as well as in attempts to improve the agricultural efficiency of the animal. Pigs are believed to be a particularly appropriate species for xenotransplantation because their organs are similar in size to human organs; they also breed rapidly, so that they could be used to supply transplant material on a large scale. In the UK, the Nuffield Council on Bioethics, an independent body comprising doctors, scientists, philosophers, theologians and others, was established in 1991 to consider the ethical issues raised by developments in medicine and biology. The Council have concluded that in a society in which the breeding of pigs for food and clothing is accepted, their use for life-saving medical procedures should also be acceptable.

As far as agriculture is concerned, there are various opportunities for the production of transgenics, especially in areas such as improved disease resistance, which would enhance their welfare. However, it is a matter of developing precise methods of gene insertion and in controlling their expression. As noted by Webb (1995), an alternative strategy may be in attempting to control the expression of existing genes by understanding the mechanisms involved.

1.3. Factors Affecting Reproductive Efficiency

In view of the fact that the profitability of the pig enterprise is largely a matter of the number of pigs weaned per sow annually, techniques, whether they are controlled reproduction procedures or otherwise, that may improve conception rates, reduce the incidence of embryonic mortality, decrease the farrowing interval and improve the rate of piglet survival are likely to be of considerable commercial interest. The number of pigs weaned, or sold, per sow annually is the most important measure of sow productivity and has a large effect on overall herd profitability; production efficiency is also influenced by the age structure of the sow herd and by the patterns of culling. As large, efficiently managed, pig units have developed in different countries, accurate observations and recording systems have provided data on reproductive efficiency.

An efficient level of reproduction in present-day breeding units means a sow producing 2–2.4 litters of young annually and rearing 18–24 pigs in that period; this is still some way short of the known biological limits of the species, yet some way in excess of the level of efficiency attained by many pig producers. Gerrits *et al.* (1979) suggested that reproductive efficiency in the USA had remained essentially unchanged during the 1970s; the figures quoted showed an average of 7.2 pigs marketed per litter, with each sow farrowing about 1.8 litters per year. A reasonable objective, these authors concluded, would be two litters per sow annually and 10 pigs per litter marketed. This was some way below what the pig is capable of producing; a mature sow herd has the potential of reproducing at the rate of 2.6 normal-sized litters per year.

According to Cunha (1980), again dealing with the North American scene, the goal for the year 2000 should be 24 pigs raised per dam per year, with producers being able to wean 90–95% of all the pigs born. In the mid-1980s, Bichard and David (1986) refer to a normal level of production in the USA of 7–8 pigs weaned per litter and 12–14 weaned per sow per year; in the

UK at that time, MLC figures recorded sows weaning 9.2 pigs per litter and 21.8 pigs per sow per year.

The rewards for achieving higher output per sow are generally clear; food constitutes 70–80% of the costs of pig production and the amount of food consumed by a sow in a year is much the same whether she rears 12 or 24 young per year (English *et al.*, 1984). For many producers, it may well be that the only way to stay in profit in pigs is by way of larger and more frequent litters and a reduction in piglet mortality; controlled reproduction techniques may be of value in achieving worthwhile improvements in these different aspects. It should also be noted that it is the degree of fertility which is of greatest interest in pigs; as in cattle and sheep, total failure to reproduce is rare and a problem of minor practical importance in the sow.

1.3.1. Early development of the pig conceptus

Fertilization

The process of fertilization in the pig has been described by Hunter (1991) in a review covering ovulation, transport of oocytes and sperm and interactions between the male and female gametes. Mating in the sow normally precedes ovulation by several hours and a reservoir of sperm is established at the uterotubal junction. The sperm move up to the site of fertilization in the ampulla at the time of ovulation; relatively few sperm are in the ampulla at any one time. It is important that sperm do not arrive in the ampulla to be faced with an aged oocyte. It is not difficult to arrange sperm and oocyte events in the pig to give rise to polyspermy.

It is evident from several reports that the development of the pig conceptus to day 11 of pregnancy (day 0 = onset of oestrus) contributes to embryo mortality by creating diversity of embryo development within litters. Morphological and biochemical hetereogenity in follicular development is known to be a feature of the peri-oestrous period in pigs and there is the possibility that embryonic diversity and embryo mortality may be the result of this follicular hetereogeneity. Such considerations led Blair *et al.* (1994a) in Kansas to examine variations in the endocrine environment during the peri-oestrous period that might be related to the events of early pregnancy. They found that pigs with high embryo survival on day 11 of gestation tended to exhibit less variation in embryonic development than did pigs with low embryo survival. It was evident that a closer synchrony of endocrine events (peak oestradiol concentration/onset of LH surge) with the onset of oestrus was associated with less embryo diversity and higher embryo survival in early pregnancy. In Denmark, Hyttel *et al.* (1994) also draw attention to the fact that various reports have shown evidence of a relationship between embryo diversity (variation in the rate of development in embryos within the same litter) and early embryo mortality due to non-infectious causes; they suggested that embryos with the slowest rate of development may be eliminated as a consequence of changes in the uterine environment triggered by the faster developing embryos.

In this context, it is necessary to stress that pregnancy in the sow, as in other farm mammals, is dependent on an appropriate relationship being established between mother and embryos. Initially, there needs to be a sequence of uterine secretions to support embryo development, followed by signals from the embryos, which are essential for establishing pregnancy. Furthermore, these changes need to occur in synchrony otherwise embryo survival may well be prejudiced.

The relationship of length of the uterus in prepubertal pigs to potential litter size was demonstrated by Wu and Dziuk (1995). They measured the length of the uterus at various ages (20–100 days); a significant correlation between number of live fetuses and fetal survival and uterine length in pregnant gilts was established. The authors suggested that prepubertal length gives an indication of postpubertal uterus length and potential litter size in pigs.

Stress and embryo mortality in sows

A study by Liptrap *et al.* (1993) in Canada investigated the effects of a hormone associated with stress (adrenocorticotrophic hormone, ACTH) on preovulatory oocytes and their development after fertilization *in vitro*; ultrastructural features of the oocytes and resultant embryos suggested that stress hormones interrupted oocyte maturation by disrupting the integrity of the vitelline membrane and by interfering with the normal action of the cortical granules (Fig. 1.2). The reduced fertilization rate and retarded growth of embryos recovered from experimentally stressed sows indicated that the developmental potential of the oocytes was compromised by the stress hormone. This may be some indication of events that occur in the reproductive tract of sows exposed to certain forms of stress. Other work in the same laboratory in Canada suggested that stress in the sow may result in a decrease in circulating oestradiol levels (Liptrap and Viveiros, 1993).

In Sweden, Tsuma *et al.* (1994) also examined the question of stress in early pregnancy as a factor influencing embryo mortality; feed was withheld on days 10 and 11 of pregnancy for 48 h. When sows were slaughtered on day 17, the percentage of embryos surviving was 64.5 and 71.4, respectively, for the fasted and control animals.

1.3.2. Genetic and environmental factors

Genetics and reproductive efficiency

Pig geneticists have considered a number of reproductive traits, both in females and males, including age at puberty, litter size at birth and weaning and sow productivity as a measure of both reproduction and mothering ability. These traits, as they have been measured in the past, have been of low to moderate heritability and have not responded well to selection. According to Rothschild

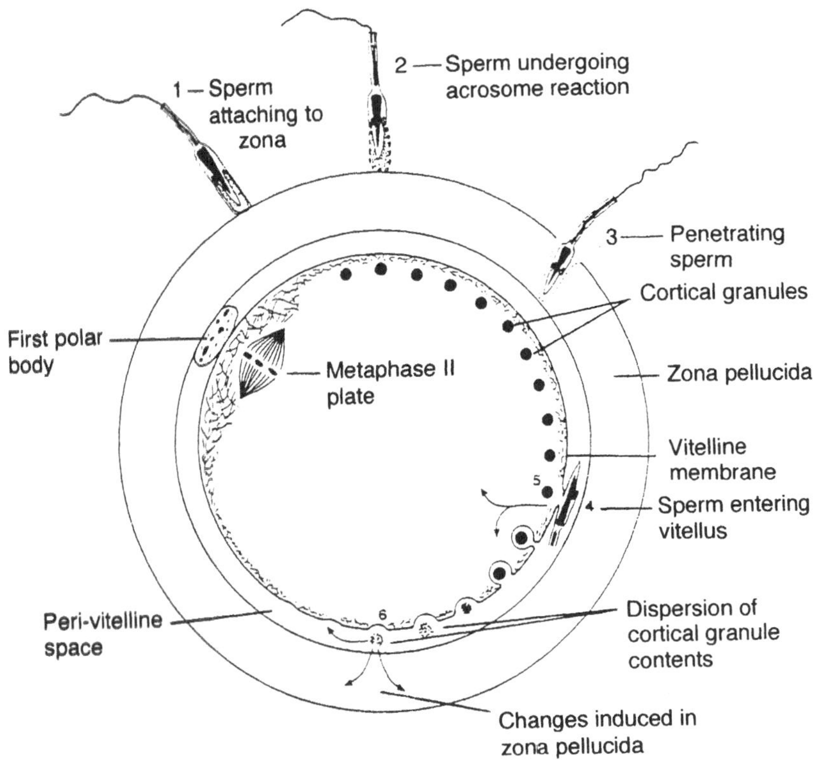

Fig. 1.2. Diagrammatic representation of events occurring at the time of fertilization in the pig oocyte.

(1990) a number of new approaches have been suggested as a way of achieving faster progress in the improvement of these reproducive traits; these include considering litter size to be made up of the component traits of ovulation rate and embryo survival or ovulation rate, embryo survival and uterine capacity. However, the author concludes that much more information about these parameters and the genetic control of hormones is required before it is possible to predict what impact these approaches may have towards genetic improvement.

Kerr (1995) notes that a model for litter size in the pig had been developed at the US Meat Animal Research Centre, Nebraska, which incorporated uterine capacity with ovulation rate and fetal survival; it was believed that as ovulation rate, fetal survival and uterine capacity interact to determine litter size at birth, a single model combining these traits may provide the most efficient means of selecting for litter size.

There is also the issue of selecting for improved male reproductive traits as a means of improving the reproductive efficiency of the sow. Testes weight and size have been measured and such traits appear to be moderately heritable and at one time were believed to be favourably associated with female reproductive traits. However, work in Nebraska where testis weight was used as an indicator for female reproductive traits apparently failed to substantiate this belief (Kerr, 1995). It was found, not surprisingly, that selection for increased testis size could be used to increase reproductive efficiency in boars, through the production of more high quality sperm. There may be other traits, such as follicle stimulating hormone (FSH) level, measurable in boars, which may yet prove to be a valuable predictor for litter traits in their female progeny.

Breed differences in performance

In countries where purebred pigs are used in commercial units, many authors have compared the performance of different breeds. In Nigeria, for example, Omeke (1989) reported a comparison of seasonal effects on Landrace and Large White sows in the subhumid tropical environment of that country; fertility traits did not differ significantly between the breeds due to season. In China, Yen *et al.* (1991) compared Landrace and Duroc sows; within parities, the total numbers of piglets born and live piglets per litter and litter weight at birth, did not differ significantly between the breeds.

Influence of lactation length

As a factor influencing the reproductive efficiency of the pig, the duration of lactation is important. An analysis of 14,925 farrowing records from 39 herds in the USA was made by Xue *et al.* (1993). Longer lactation lengths were associated with significantly higher subsequent litter sizes (both total born and born alive), shorter weaning-to-service intervals, longer farrowing-to-service intervals and longer farrowing-to-farrowing intervals. The number of live pigs born per sow annually was not significantly associated with length of lactation. The authors showed that reducing the duration of lactation from 30 to 17 days did not significantly affect the number of pigs born per sow annually; this was attributed to the compensatory effects of the shorter lactation on pigs born per litter and litters per sow per year.

1.3.3. Reproduction and the lean pig

Bichard (1980) pointed out that, in making the modern bacon pig so much leaner, the sows of today are probably not so well buffered against fluctuating energy demands as those of yesteryear (Fig. 1.3). For that reason, lean meat sows may require a higher standard of management if they are to be kept at peak performance throughout their lives. Responses in reproductive performance have shown that selection for particular aspects of efficient lean growth adversely affects reproductive performance in Large White pigs (Kerr and Cameron, 1994). However, such negative effects of selection strategies for lean growth may be reduced by incorporating reproductive traits in the selection objective (Kerr and Cameron, 1995). These authors suggest that alternative breeding strategies, which take due account of responses in reproductive

Fig. 1.3. The lean meat, highly productive, pig of today may not always have the reserves to cope with efficient reproduction under some management conditions.

performance, need to be evaluated to optimize progress in genetic improvement in lean meat production.

1.3.4. Nutrition and sow reproduction

Within the pig, the internal environment is held relatively constant by homoeostatic mechanisms that involve a complex interaction of biochemical signals. Inputs from the external environment can influence the delicate balance of such signals. In quantitative terms, the most important external factors are nutrition, ambient temperature, light and infectious agents. (Fig. 1.4). Feeding strategies which aim to minimize weight loss and maintain body condition in the sow throughout the various stages of reproduction have been discussed by various authors (Aherne and Williams, 1992; King, 1993); sows should be fed freely from day 1 after farrowing until weaning.

Reduced feed intake during lactation is likely to lead to a loss in weight and condition which will extend re-mating intervals, and result in a reduced pregnancy rate and reduced embryo survival (Zak *et al.*, 1995). The optimum feeding strategy to ensure minimum non-productive days in the sow should involve a carefully controlled and limited weight gain during pregnancy, followed by maximum conservation of body tissues during lactation.

On the other hand, there is evidence supporting the view that high-level

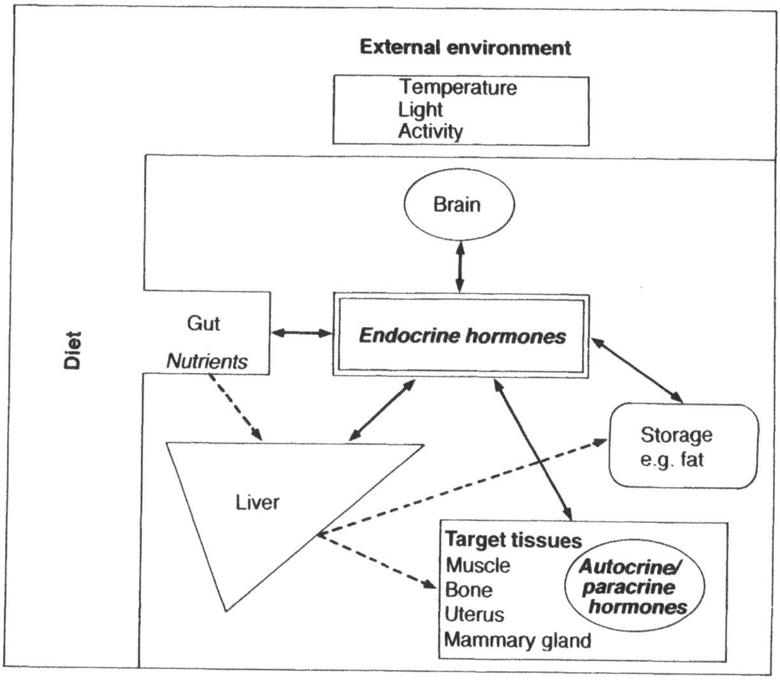

Fig. 1.4. Interrelationship of external and internal factors involved in the maintenance of homoeostasis in the pig (from Gilmour *et al.*, 1992).

feeding in early pregnancy may result in an increase in embryo mortality in gilts (Kirkwood and Thacker, 1988); there have also been studies in gilts showing that the 25–30% reduction in embryo survival arising from high-level feeding could be avoided by the administration of exogenous progesterone (Ashworth, 1990). As with sheep, and to some extent cattle, it would appear that high food intakes, as a consequence of their stimulatory effects on hepatic blood flow and the rate of progesterone clearance in the liver (Symonds and Prime, 1989), may result in a decrease in plasma progesterone to the extent that embryo survival is compromised.

The relationship between feeding and progesterone levels and between progesterone levels and embryo survival have been examined by a number of workers. Data presented by Cassar *et al.* (1994) at Guelph, indicated that in adequately fed gilts, increased dietary energy after mating does not affect conceptus survival during early gestation. Elsewhere in Canada, Jindal and Foxcroft (1995) have shown that a 2× rather than a 1.5× maintenance ration per day from day 1 after the onset of oestrus resulted in a significantly higher embryo survival rate (85% vs. 64%) and increased plasma progesterone concentration (10.5 vs. 4.5 ng ml^{-1}); the study substantiated the hypothesis that nutritional effects on embryo survival are probably mediated by differences in progesterone concentrations.

The effect of nutrition on sow reproduction has been reviewed by Pettigrew and Tokach (1991) who noted that excessive energy intake during pregnancy reduces feed intake during lactation and appears to impede mammary development. There are indications that protein restriction during lactation may be more detrimental to subsequent reproduction than energy restriction.

Although the relationship between nutrition and reproduction has been observed in many pig trials, the mechanisms by which the two are linked remain obscure. In Ireland, Lynch *et al.* (1995) examined the effects of supplemental fat in the diet on follicle development, ovulation rate and embryo survival; although the addition of lipid to the sow diet did not influence ovulation rate, the inclusion of 5% tallow in the diet did increase the number of viable embryos present at day 30 of gestation.

Vitamins and litter size

Adequate dietary levels of several vitamins are important for maximum reproductive performance in the sow (Pettigrew and Tokach, 1991). The enhancement of sow reproductive performance by intramuscular injection of β-carotene or vitamin A was reported by Coffey and Britt (1993) in North Carolina; they recorded that the number of piglets born alive was significantly increased in sows fed diets that were adequately fortified with supplemental retinol. Other studies in the USA also demonstrated that the injection of vitamin A six days before mating increased embryonic survival to day 11 or 12 of pregnancy and litter size. Data reported by Whaley *et al.* (1994) suggested that treatment with vitamin A before ovulation may influence the follicular environment, thereby synchronizing resumption of meiosis and enhancing early embryonic survival in pigs.

Adverse nutritional factors

The effects on sow reproduction of antinutritional factors such as mycotoxins (zearalenone) and glucosinolates (from rapeseed meal) have been reviewed by Etienne and Dourmad (1994). Unlike nutrient imbalances where effects may appear late and prove difficult to cure, the influence of such antinutritional factors may be immediately obvious and disappear when their intake is stopped. Zearalenone, which is produced by some strains of *Fusarium* fungi, contaminates cereals harvested in damp conditions; it is characterized by its oestrogenic properties which induce vulvovaginitis in gilts, anoestrus in cyclic females or delayed return to oestrus after the piglets have been weaned. In the pregnant animal, zearalenone reduces embryonic survival when given beyond a certain threshold and may decrease the weight of the fetus.

Rapeseed meal contains glucosinolates whose metabolites have antithyroid properties; in the pregnant sow, they may induce thyroid hypertrophy in both mother and fetuses. It is believed that pig fetuses may be especially sensitive since their plasma thyroxine and liveweight decrease in relation to glucosinolate in the sow's diet; harmful effects should not occur if dietary glucosinolate levels remain below 2 $\mu M\ g^{-1}$. A report by Love *et al.* (1990) in Australia has shown that increasing the level of cottonseed meal in sow diets from <5% to 10% significantly increased the incidence of premature farrowings from 1.1% to 2.7% and significantly decreased the mean gestation length; after removal of the cottonseed meal from the diet, there was a residual effect lasting several weeks before the gestation length returned to normal. The mechanisms involved in this are still not understood.

1.3.5. The Meishan breed in research and commercial practice

As noted earlier, it is now well established that the purebred Meishan has a litter-size advantage of three to four piglets per litter; the litter-size advantage is believed to stem from better embryo survival and a higher ovulation rate (Table 1.2). It should also be noted that the Meishan has an average of three more teats than the Large White, so that in postnatal life, competition for teats among piglets would not be expected to increase significantly.

Precocious reproduction may be useful in certain areas of research in reproductive physiology and endocrinology. Prepubertal pigs may have certain advantages as experimental animals because they can be used without reference to an oestrous stage and are likely to be easier to handle due to their small size. More than 60 years ago, Casida (1935) showed that follicular development and ovulation could be induced with exogenous gonadotrophins in prepubertal pigs. Numerous workers have used pregnant mare serum gonadotrophin (PMSG) alone or in combination with human chorionic gonadotrophin (hCG) to induce follicular development and ovulation in prepubertal pigs to obtain oocytes, zygotes and embryos. Although ovulation cannot be induced by exogenous gonadotrophins until 100 days of age in European and American breeds (Dial *et al.*, 1984), in the Chinese Meishan, ovulation can be induced in 33% of pigs at 60 days of age and in all pigs at 75 days (Miyano *et al.*, 1990).

The need to increase the efficiency of production of lean meat is one of the greatest challenges facing pig producers, and Chinese breeds, such as the Meishan, offer a valuable resource to increase prolificacy in pig-breeding populations and to conduct basic research on the reproductive biology of the pig. If the current efforts to incorporate the prolificacy of the Meishan without

Table 1.2. Comparative data for Meishan, Large White and their crosses from studies in the UK (from Webb, 1995).

Trait	Large White	Meishan	F_1 cross
Born alive	9.8	13.2	13.8
Ovulation rate	14.9	18.9	17.7
Prenatal survival (%)	66	71	78
Piglet birthweight (kg)	1.28	0.93	1.20
Number of teats	14.2	17.3	16.3

loss in growth or carcass quality over currently produced market pigs are successful, the improvement in efficiency of production could be considerable.

There have been many studies dealing with factors responsible for the extreme prolificacy of the Meishan, in the UK (Haley, 1991; Biggs *et al.*, 1993; Hunter *et al.*, 1993, 1994a,b; Faillace and Hunter, 1994; Haley *et al.*, 1995; Lee *et al.*, 1995; Lee and Haley, 1995; Sinclair *et al.*, 1996; Southwood *et al.*, 1996), the USA (Gerfen *et al.*, 1992; Youngs *et al.*, 1994; Ford, 1995; Macdonald *et al.*, 1995; White and Wheeler, 1995; Wilson *et al.*, 1995), and in France (Terqui *et al.*, 1992; Prunier *et al.*, 1993; Downey and Driancourt, 1994). It is apparent that as well as showing a higher ovulation rate than indigenous European breeds, the Meishan apparently experiences less embryo mortality between days 11 and 20 of pregnancy than indigenous breeds.

Ovulation rate in Meishans

A report on periovulatory hormone profiles in Meishan and Large White sows (Hunter *et al.*, 1995) suggested that the higher ovulation rate in Meishans was associated with enhanced oestradiol and inhibin secretion and a more rapid luteinization response to the LH surge. At Nottingham, where previous studies had shown that Meishan sows have a higher ovulation rate than Large White pigs, a report by Miller *et al.* (1995) showed that the dynamics of follicle development differs between the breeds. It appeared from this report that the Meishan sow possessed a larger pool of ovarian follicles which was maintained later in the follicular phase; it was speculated that this may represent a mechanism by which the Chinese breed achieves its higher ovulation rate. Other work at Nottingham has also led to the suggestion that advanced oocyte maturation in the follicles of Meishan pigs before ovulation may be important for ensuring the prolificacy of this breed (Faillace and Hunter, 1994).

Oestradiol secretion

In Aberdeen, Pickard and Ashworth (1995) investigated the timing and control of oestradiol secretion by the developing pig embryo. They concluded that aromatase enzyme activity controlled the onset of oestradiol secretion by Meishan embryos, which appeared to be more synchronous in the Chinese breed than in the Large White. The Rowett authors suggest that this may contribute to enhanced Meishan conceptus survival. Other work in Aberdeen included studies comparing endometrial and conceptus protein secretion between Large White hybrids and Meishans (Ashworth *et al.*, 1995) and an examination of possible ways by which Meishan sows are able to limit fetal growth (Ashworth and Atkinson, 1995).

Synthetic pig strains with Meishan blood

Pigs with Meishan genes have already appeared on the UK market (Webb, 1995). At the Cotswold Pig Development Company, 25% Meishan parent gilts have shown an advantage of one to two extra live piglets per litter; the potential extra fat in slaughter progeny can be controlled by mating to an ultra-lean terminal sire and/or restricting nutrient intake.

Characteristics of Meishan boars

Early puberty is evident in Meishan boars. Harayama and Kato (1993) in Japan found that Meishan boars acquired the ability to ejaculate at about 2 months of age, although they were not suitable for mating until a later age.

1.3.6. Fertility and temperature

The condition known as summer infertility, characterized by a decline in the pig's reproductive performance during hot weather, has been recognized by pig producers and veterinarians in many parts of the world. Aumaitre *et al.* (1976) in France found that sows farrowing from July to September had a 10-day longer weaning-to-conception interval than sows farrowing at other times of the year. In the USA, several reports also showed that the farrowing interval could be markedly prolonged for sows farrowing in the June to August period, compared with other months of the year (Hurtgen and Leman, 1981); it was also evident that primiparous sows were more prone to summer infertility than older animals.

In the UK, in a study of several large pig herds, Stork (1979) found that reproductive efficiency was related to temperatures in excess of 20°C; in practical terms, pigs mated in July and August showed an increase in returns to service. On the other hand, in the climate of west Scotland, in which the cool summer months offer little chance for heat stress to occur, evidence of seasonal variation in reproductive efficiency was not recorded by Pepper and Taylor (1981) in an analysis of several thousand farrowings throughout a 12-month period. In general, annual sow reproductive performance differs between countries in colder climates and those in the tropics where annual output has been estimated to be 15–20% lower; such differences persisted even when similar breed and husbandry methods were employed.

In Australia, the annual output of pigs per breeding sow has been recorded as substantially below that achieved in some of the European countries. In the 1970s, Australian workers estimated that Australian sows produced about 3.4 pigs less per year than those in the UK, the difference being mainly the result of small litters, which were almost two piglets below the average litter size in the UK. Within Europe itself, the number of pigs slaughtered per sow per year was recorded as being much higher in the northern European countries than those in the south (King, 1970).

The effects of season on semen traits of boars in the tropics was studied by Fuentes *et al.* (1993) who concluded that high ambient temperature has an adverse effect on the quality and production of sperm.

Experimentally elevated temperatures

On the research front, experimentally elevated ambient temperatures have been shown to increase the incidence of anoestrus in gilts and reduce conception rate and embryo survival in others. Many of these investigations led to the conclusion that the effect of heat stress was on the sow's fertility but there were some who suggested that high temperatures might stress the boar and reduce semen quality and male fertility. However, there does not appear to be clear evidence that semen quality in boars is markedly affected by season of the year (Cameron, 1980; Hurtgen *et al.*, 1980).

1.3.7. Sow culling strategies

Pattison (1980) drew attention to the low breeding-life expectancy of sows in many commercial herds, resulting from a high culling rate (30–50% in the early parities) and the fact that such high culling levels could involve substantial economic loss. If the average age at culling was somewhere between the second and fourth litters, this would mean that a high proportion of sows would be culled before they reach their peak in terms of litter size. This would obviously be to the detriment of overall herd performance. In France in the early 1980s, Martinat-Botte *et al.* (1985) recorded a 40% culling rate; in practical terms, for a unit of 100 sows, this meant that about 120 gilts must be purchased or home-bred each year (at 2.3 farrowings per year).

Whether the culling of sows for reproductive problems at levels of 30–50% would always be justified is another question. In France, where workers examined several hundred breeding tracts from sows culled by producers, they concluded that about 40% of the pigs had been culled for unjustified reasons. It appeared that culling errors often occurred as a consequence of faulty oestrus detection, especially in herds where boars were housed independently of sows and where the stockperson attempted to detect heats in the absence of a boar.

Data reviewed by Dourmad *et al.* (1994) in France showed that sows are culled mainly for reproductive failure, locomotion problems or poor lactation performance. Overfeeding during pregnancy can increase weight and condition of the sow at the end of pregnancy with effects on the occurrence of farrowing and lactation problems; this may lead to culling for poor lactation performance or locomotion. The French authors note that during lactation there are critical levels for energy and protein intake below which weaning to oestrus interval, rate of return to oestrus and pregnancy rate are affected, especially in primiparous sows; this can be the main reason for the early culling of sows. Although the mechanisms are not clear, it appears that the metabolic status or the status of body reserves of the sow at weaning may be involved. It is believed that it may be necessary to apply a feeding strategy adapted to each sow, in order to maintain its body reserves within an optimal zone throughout reproductive life and to maximize longevity.

1.3.8. Stress, records and reproduction

Much research effort is currently expended in evaluating areas of management that may give rise to stress in pigs and consequently exert an adverse effect on their reproductive performance. Varley (1991) has reviewed work on the effects of stress arising from management practices, the environment or nutrition on reproduction in sows and gilts. Some forms of stress may not be immediately obvious and may require a detailed study of performance records before they are revealed. In Australia, for example, it was found that sows mated early in the week achieved a significantly higher farrowing rate than animals mated late in the week (Love and Wilson, 1990). An analysis of farrowing rates on a day of mating basis led to the conclusion that the stress of regrouping unmated sows on a Wednesday to accommodate newly weaned sows was probably responsible for the decline in fertility.

1.3.9. Effect of antibiotics and hormones

Chlortetracycline treatment

A study reported by Maxwell *et al.* (1994) dealt with 850 litters in several states in the USA involved in assessing the effects of feeding chlortetracycline (CTC, 200 mg kg^{-1}) from one week before mating to 15 days after mating (breeding period) and/or from 110 days of gestation and throughout lactation on the reproductive performance of sows. Feeding with CTC during the mating period had no effect on conception rate; however, during lactation, CTC significantly improved subsequent conception rate at the first service (80% vs. 73%). The results indicated that feeding with a low concentration of CTC (1.2g day^{-1}) improved overall reproductive performance.

1.4. Factors Affecting Semen Quality and Sex Drive in Boars

1.4.1. Aspects of boar fertility

Reproductive organs of the boar

The scrotum of the boar is just below the anus and is not so well defined as it is in farm ruminants and horses (Fig. 1.5). The testes are large but relatively soft and are almost horizontal in the scrotum. The sperm duct (ductus deferens) lacks a distinct ampulla and the vesicular and Cowper's glands are large and dense; the bulbourethral (Cowper's) glands are nearly cylindrical and lie astride the pelvic part of the urethra. The penis is fibroelastic and its distal free end is corkscrew-shaped. The sigmoid flexure is located anterior to the scrotum. The prepuce has a narrow orifice with stiff hairs and in the upper wall of the preputial cavity is an opening into an ovoid pouch, the preputial diverticulum. The preputial pouch contains a mixture of decomposing urine and macerated epithelial cells which has a characteristic and very unpleasant odour. This is believed to be responsible for the strong sex odour of the boar, which permeates its meat to give it a disagreeable taste.

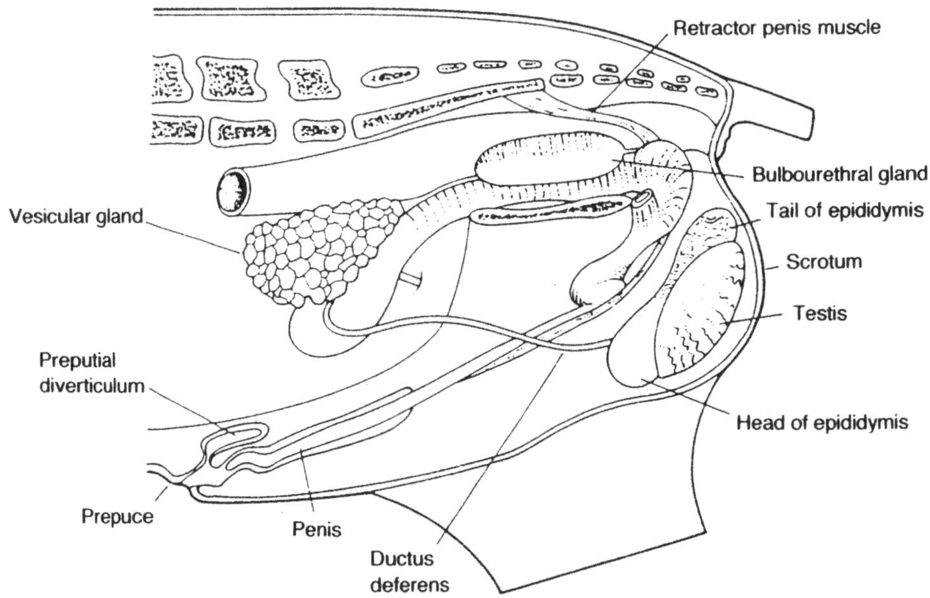

Fig. 1.5. Diagrammatic representation of the boar's reproductive tract (after Ashdown and Hancock, 1974).

Morphology of the epididymis

The seminiferous tubules terminate in the rete testis from which 10–12 efferent ducts join to form the epididymal duct proper, which is a single tortous duct with an average length of 63 m; it is lined by pseudostratified epithelium. On the basis of the epithelial morphology, the boar epididymis can be divided into several regions comprising three functional parts; initial, middle and terminal segments (Glover and Nicander, 1971). The locations of these segments are shown in Fig. 1.6.

Boar sperm undergo physiological, morphologial and biochemical changes as they pass through the epididymal duct; transit time for sperm is estimated to be two weeks. It is believed that sperm can then be maintained in the terminal segment of the epididymis for a prolonged but unknown period and may ultimately be excreted with the urine if the boar is sexually inactive. Sperm acquire the ability to fertilize in the middle segment of the epididymis and increasingly greater proportions of sperm possess this ability the further distal in the duct they progress (Crabo, 1985).

It is known that macromolecules of importance for the fertilizing ability and motility of boar sperm are acquired in the initial segment of the epididymis; these membrane-bound molecules are then covered with peripheral proteins in more distal segments of the epididymis.

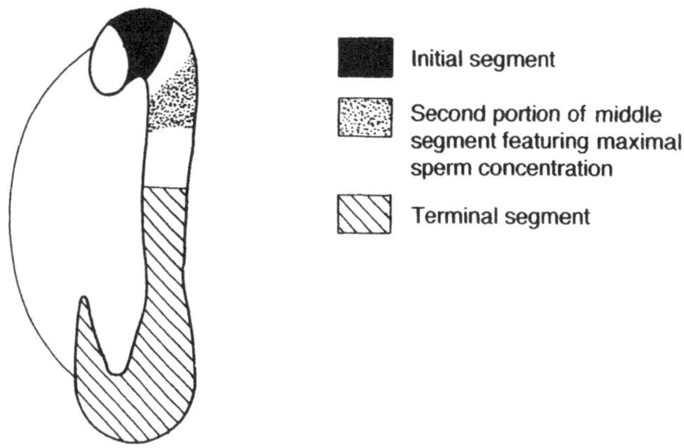

Fig. 1.6. The functional topography of the boar epididymis (from Crabo, 1985).

1.4.2. Effect of nutritional and environmental factors

Nutritional strategies

A review of energy and protein requirements for the boar (Kemp and Hartog, 1990) concluded that protein intake influenced semen quantity but had little effect on the semen quality or libido of the boar; a certain minimum protein intake is apparently necessary for a good sperm count, but reports do not agree on what this intake should be. The authors suggest that in practice a commercial sow diet (containing, per kg: metabolizable energy, 12.56 MJ; crude protein, 145g; lysine, 6.8g; methionine + cystine, 4.4g) should provide sufficient protein to ensure good reproductive performance in the boar. Feeding level or energy intake were not found to affect semen quality and high mating frequencies had little affect on the feeding requirements of boars.

Similar recommendations were contained in a further review by Kemp (1991). In this, the author recommends that an extra allowance of 100 g of feed per degree below an ambient temperature of 20°C should be provided; if this was not given, it may result in reduced semen production. In the USA, Louis *et al.* (1992) examined the effect of energy and protein intakes on boar libido and semen characteristics; when protein and energy intake was low, boars either consistently refused to mount the dummy or produced less semen than boars with a high protein intake.

In terms of beneficial effects arising from the inclusion of specific ingredients of the boar diet, Brzezinska-Slebodzinska *et al.* (1995) in Poland reported that vitamin E supplementation significantly increased the sperm concentration in the ejaculates of boars. They were able to demonstrate the protective effect of vitamin E and glutathione against fatty acid peroxidation of boar semen and the benefit of vitamin E supplementation on semen quality.

1.4.3. Space requirements

A survey of pig breeding herds in Scotland by Petchey and Hunt (1990) showed that most matings (72%) in that country took place in the boar pen; the authors drew attention to the fact that a small pen inhibits mating because there is insufficent space for the animals to manoeuvre. The authors concluded that a boar pen is too small to act as a mating arena when it does not allow the boar and sow to stand in line nose to tail.

1.4.4. Light programmes and boar reproduction

There have been studies in pigs which suggest that photoperiod may affect growth and reproductive performance. Supplemental lighting has been shown to hasten puberty in the boar (Hoagland and Diekman, 1982). In Germany, Weiler *et al.* (1988) examined the effect of light programmes on boars kept for artificial insemination (AI); they recorded that an abrupt decrease in daylength was effective in stimulating reproductive function in the males. In the former East Germany, Mudra *et al.* (1990) subjected boars to a constant daylength of 8.5h or to natural daylength (controls) over a period of a year. Boars on natural daylength had significantly larger ejaculate volumes than those on light control, but there were no significant differences between the two groups in sperm density, motility or viability or in the incidence of sperm abnormalities; sows inseminated with semen from the two groups did not show any significant differences in pregnancy rate and litter size.

1.4.5. Fertility problems in the boar

The boar has an important role in determining the reproductive efficiency of the breeding herd (Fig. 1.7). Reasons for infertility in boars include reduced libido, failure to mount, impotence, poor semen quality and defects of the genitalia (Dirks, 1989; Potter, 1995). In commercial pig farming, objective methods whereby the reproductive performance of the boar can be evaluated and assessed are required. Some Australian studies reported in the early 1980s emphasized the important part that stimulation from the female pig has on the sexual behaviour of the boar. It was evident from such studies that boars isolated from female pigs often achieved fewer matings and displayed less courting behaviour than boars housed near sexually receptive sows; the practical lesson in that was that those operating pig AI centres could optimize the sexual behaviour of their breeding boars as well as intensifying the use of genetically superior males by housing the boars near female pigs.

Fig. 1.7. Factors influencing the reproductive performance of the boar are important in determining herd efficiency.

1.4.6. Evaluating boar sexual behaviour

The differentiation of sexual behaviour in the young boar has been discussed by Signoret (1989). It is believed that sexual behaviour increases during the second month of life and is not related to endocrine changes. Approaching puberty, which occurs from three months of age, boars show bisexual behaviour, mounting oestrous females and becoming immobile when mounted by an adult boar. Differentiation of sexual behaviour is slow and is incomplete at puberty.

According to Hemsworth and Winfield (1986), experience from pig AI centres and commercial pig units suggests that some 20–50% of culled boar are unable to copulate or copulate at an insufficient frequency; the condition stems from either low sexual motivation or low mating competency. There are indications that low copulatory capacity is more commonplace in intensive than extensive production systems and that its incidence increases with age, perhaps because of increased susceptibility to locomotor problems in such systems. These authors note that after 24h of rest, a boar of adequate sexual motivation, when introduced to an oestrous sow or gilt, should quickly engage in courtship and attempt to mount within two minutes; once the boar copulates, ejaculation should last for 5 to 15 minutes.

Attempts to evaluate sexual behaviour of boars on some objective basis have been reported in some studies. In the USA, Levis *et al.* (1989) evaluated response over a 5-minute period to a tethered, ovariectomized, oestrus-induced gilt; boars were classified as having low, medium or high levels of sexual behaviour. On the basis of an index calculated from their performance in the test, the subsequent mating behaviour of boars was reliable in those classified as having high or low sexual behaviour.

1.4.7. Boar culling rates

Studies examining the boar culling in 84 breeding herds in Minnesota showed the annual culling rate averaged 59% (D'Allaire and Leman, 1990). Removal of the boar was the result of being overweight (47%), reproductive problems (18%), leg problems (12%), death (7%) and other diseases (4%). From the annual culling rate, the average breeding life of boars was estimated to be 20 months.

1.4.8. Efficiency of the mating process

The critical importance of the mating process to the success of any pig-breeding programme requires no emphasis. It is, however, useful to examine ways whereby the quality of the boar's service may be assessed. Workers at Aberdeen examined the mating process to determine potential sources of inefficiency (Rickard-Bell *et al.*, 1993); they graded the quality of service according to the scoring system outlined in Table 1.3. Some sows stood very firmly during mating and grunted with apparent content whereas at the other extreme sows were very restless with vocalizations indicating discontent. There was a significant correlation between the duration of intromission and subsequent litter size; quality of service was related positively to litter size but not to pregnancy rate.

Number of services

In breeding pigs under normal farm conditions, there is a need for accurate heat detection to ensure that all the females are bred on the chosen date. Because the heat period may last for two or even three days, and ovulation is known to occur about 40 h or so after its onset, it is also important to know the optimum time for breeding (natural service or artificial insemination) within the heat period. If high fertilization rates are to be achieved, the general view is that sperm should be in the oviducts about 12 h before ovulation. This was shown some time ago by inseminating sows at known intervals before and after ovulation (Hunter and Dziuk, 1968).

Although a single service at the optimum time during oestrus should be sufficient to ensure a high fertilization rate, in practice, the observation of heat onset is seldom accurate enough to enable service to be delayed with confidence until the second day of oestrus. The usual policy is to try to ensure that sows are served once on each day that they stand for the boar (Rickard-Bell

Table 1.3. Criteria used in scoring the quality grade of first services of sows (from Rickard-Bell *et al.*, 1993).

	Grade			
Criteria	1 (Poor)	2	3	4 (Excellent)
Number of attempted mounts before intromission	Several mounts before intromission or a series of short intromissions	More than one mount before intromission or a series of short intromissions	Single successful mount	Single successful mount
Apparent comfort/agitation/ annoyance during mating based on sow vocalizations	Agitated/annoyed	Agitated/annoyed part of time	Contented	Contented
Restlessness of the sow during mating	Moving insecurely while mounted	Intermittent movement	Stood firmly but slight movement towards end of copulation	Stood firmly throughout mating
Duration of intromission (s)	<90	90–180	>180	>180

et al., 1993). In the majority of pigs, this means two matings, but for those sows that are not detected until the second day, a high conception rate would still be ensured. Herds in which sows routinely receive two services per heat period have been found to have conception rates 5–10% higher, and also a tendency for larger litter sizes, than those which receive only the one service.

However, there have been some who question the need for multiple services for sows and lay much more emphasis on the importance of the correct timing of the first service. It has been speculated that semen from the first service may adversely affect the entry of semen from the second mating. However, if there is one service, it is crucial to ensure that the single service occurs in the sow's most fertile period.

1.4.9. Overuse and underuse of the boar

Walker (1980) observed that there is evidence to show that both overworking and underworking can cause the boar to sire smaller than optimal litters. Walker (1980) also looked at data based on more than 2000 farrowings in Northern Ireland and England and concluded that resting the boar between sows increased litter size (Table 1.4). A Boar Chart should be employed to monitor the usage of males and ideally each boar should be used for one double

Table 1.4. The effect of resting boars on litter size (from Walker, 1980).

Rest period prior to mating (days)	No. of litters	Litter size
0	289	9.51
1–2	455	10.07
3–4	253	10.13
5–6	241	10.46
7–9	167	10.44
10–30	200	9.56
>30	36	9.77

service per week. Others have pointed to the loss of fertility in the boar through overuse; Rasbech (1969) suggested that young boars should not be used more than once daily and Leman and Rodeffer (1976) observed that boars with an unsatisfactory fertility record were often young males used excessively while under 8 months of age.

Hughes and Varley (1980) suggested that a boar should serve two sows per week, on the assumption that this involved double-service on each occasion; the same authors also made the point that young boars should be introduced carefully and gradually to their duties if their fertility and libido is to be maintained at a high level. It is clearly important that unsatisfactory boars should be eliminated from the breeding scene at the earliest opportunity; in this regard, there is an obvious relevance of early pregnancy diagnosis techniques that can be used in the female to show up the boar's competence.

1.4.10. *Boar stimuli at mating time*

It is known that specific boar stimuli at or around the time of mating and insemination influence sperm transport, the ovulation process and subsequent reproductive performance. A review by Soede (1993) draws attention to the way in which reproductive function in sows is affected by external factors such as olfactory and tactile stimuli and by internal factors such as seminal plasma and certain of its components. It is also evident that the social conditions of sows may alter the effects of boar stimulation on reproductive processes.

1.4.11. *Endocrine factors in boar sperm production*

Some workers have attempted to stimulate sexual behaviour in boars by way of short-term hormone treatment. In the USA, for example, Wetteman *et al.* (1992) treated boars which had been identified as lacking mating ability with prostaglandin F2-alpha ($PGF_{2\alpha}$); such treatment did not increase the number of boars that mounted or mated. Others have examined novel approaches to increasing sperm production in the boar on a long-term basis. Research in recent years has shown that neonatal hypothyroidism in adult male rats induced by a goitrogen (6-propyl-2-thiouracil) can result in reduced body-weight and fat composition, while increasing the size of the testis and sperm production.

According to Tarn *et al.* (1995), if this methodology was applicable to pig-breeding programmes, potential benefits could include decreased maintenance costs by reducing mature body size, easier boar management and increased semen production for artificial insemination. However, in pigs, variable results were obtained, with goitrogen treatment increasing testicular weight in some boars but not in others. The authors concluded that further studies were required to determine the mechanisms that result in a genotype × goitrogen interaction on testicular characteristics and sperm production.

1.5. Artificial Insemination in Pigs

Artificial insemination can be used in various ways as the method of mating in pig herds (Flowers, 1992). It may be a matter of using fresh or occasionally frozen semen from an AI centre or it may be a matter of using AI with semen produced on the sow owner's farm; it may also be a matter of using AI in combination with natural service. In Thailand, for example, where there is increasing use of pig AI, many farmers practice 'mixed mating' in which each service involves both AI and natural mating; this has been claimed to result in 0.5 extra piglets per litter compared with AI-only or natural mating-only service (Anon, 1993a). On-farm pig AI involves semen collection, dilution and insemination on the pig unit as opposed to purchasing semen from an AI centre.

1.5.1. Use of pig AI in various countries

According to Glossop (1991), on-farm AI is widespread in various European countries; in Spain, at that time, it represented 75% of AI usage. The same author noted that inseminations accounted for about 11% of sow services in the UK, the technique having been increasingly applied during the previous decade; by 1996, the figure had risen to 14% but at this level the influence of superior boars on the slaughter generation was still minimal. In France, Mani (1991b) showed that in 1990, 270,000 sows, representing 12% of the total population, were inseminated; it was envisaged that 30% of sows in France would be inseminated within the next few years. Of the 754 AI boars used in 1990, 36% were Pietrain × French Large White, 24% Pietrain, 23% commercial hybrids, 10% French Large White and 7% French Landrace. A report for 1992 (Anon, 1993b) showed that some 35% sows were inseminated in France in that year.

In Germany, figures for 1992 showed 2.2 million sows inseminated (mainly Pietrain semen) and it was evident that there was increasing use of AI; in 1993, the figure for sows inseminated had increased to 2.45 million (Gronbach, 1995), which represented about 36% of the breeding pig population. Elsewhere in Europe, Kupferschmied and Luginbuhl (1990) noted that pig AI was introduced into Switzerland in 1965 and that in 1988/89, 40,000 first inseminations (16% of breeding sow population) were carried out. Mani (1990) noted that 48% of sows in the Netherlands and 26% of those in Austria were bred by AI at that time.

Pig AI in the USA

At the start of the 1990s, it was evident that pig AI in the USA had not increased much beyond 1% of the US sow population (Crabo, 1991). For pig AI in the USA to become a viable alternative to natural service it is believed that an effective freeze–thaw procedure is essential (Schoenbeck and Didion, 1995a,b). The use of fresh semen AI in various regions of the USA has been discussed by several authors. In North Carolina, Flowers (1995) recorded that of the 140,000 pigs inseminated in 1994, less than 5% were bred exclusively with AI; the remaining 95% received an initial natural mating followed by at least one subsequent artificial insemination. The primary impetus for the use of AI in North Carolina was apparently genetic improvement of lean growth in pigs; savings in labour, boar and other costs played little part in the decision to use AI.

Pig AI in Australia

The role of AI in the Australian pig industry in the late 1980s was discussed by McIntosh (1989). A comparison of farrowing percentages of sows inseminated with semen extended with Guelph diluent and stored for 1–5 days showed that the farrowing percentage decreased from 89.5 for sows inseminated on the day of collection to 55.9 for 5-day-old semen. The total number of pigs born and the number of pigs born alive declined from 11.5 to 8.1 and 10.7 to 7.9, respectively. In terms of imported frozen semen, the author noted a farrowing rate of 45% and an average litter size of 7.6.

Problems in application of AI

Not all countries report an increasing utilization of pig AI. In Hungary, Wekerle (1993) suggested the lack of strict control of AI stations, inappropriate semen quality and other problems as the main reasons for a decline in the use of the technique in that country.

1.5.2. Collection of semen

Mounting behaviour in the boar can be released by a variety of inanimate devices ranging from dummy sows to objects that look more like a table than an artificial female (Fig. 1.8); it appears that the main stimulus to the release of mounting behaviour in the boar is a visual one. Young boars with good libido

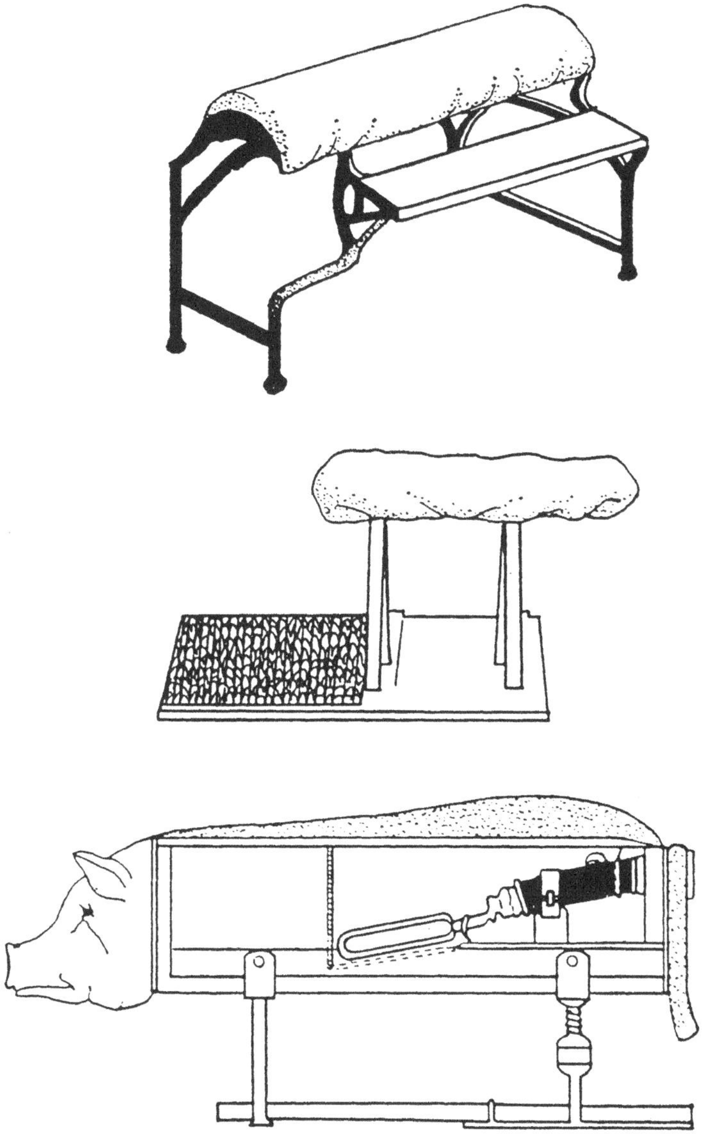

Fig. 1.8. Devices used in the collection of semen from the boar (from Signoret *et al.*, 1975).

will readily mount a dummy after observing the sexual activity of other boars on the device; if semen is collected, the response will be retained (Signoret *et al.*, 1975). Dummies have an advantage over live mounts in providing stability and preventing possible disease transmission. In the collection of semen, manual manipulation is effective for the males of species such as the pig where pressure, rather than temperature, is the main sensory stimulus for receptors in the penis.

Satisfactory collection of boar semen can normally be achieved using the 'gloved-hand' technique; in this, the gloved hand readily adapts to the cork-screw shape of the boar's penis and can provide the constant pressure necessary during the stimulatory and ejaculatory phases of the collection process. Care is required to avoid problems arising with certain types of glove. In the USA there have been reports showing the spermicidal effect of latex gloves during collection (Ko *et al.*, 1993; Wiley, 1993); vinyl gloves do not give rise to such problems.

Semen production and quality may be significantly affected by the age of the boar and by the interval between collections (Bussiere and Bariteau, 1992). Ejaculates are usually collected from boars once or twice a week and filtered to remove the gel fraction; in Argentina, Williams *et al.* (1991) concluded that under their conditions collection of semen two or three times weekly produced the best semen quantity and quality. Immediately after collection, semen should be protected from light and placed in an environment where temperature can be maintained at 30–37°C until the ejaculate can be examined and processed.

The effect of semen collection frequency (three or five times in two weeks) and food intake (high or low levels) on semen production in Large White boars was studied by Kemp *et al.* (1991) in the Netherlands, boars on a high feeding level produced more sperm than those on a low. The authors did not observe any effect of food intake or semen collection frequency on semen quality.

There may be some situations in which electroejaculation may be used. Evans and Ko (1990) have described electroejaculation in Vietnamese pot-bellied miniature pigs; in this, ejaculation was achieved by alternating 3–4 s of stimulation with 5–6 s of rest.

1.5.3. Time of insemination relative to ovulation

The effects of different times of AI relative to ovulation were investigated in Germany by Waberski *et al.* (1994b) who used real-time ultrasonics to detect ovulation in pigs inseminated with fresh or frozen semen. The authors recorded a significantly higher percentage of embryos at 28 days of gestation after preovulatory insemination than for AI 0–4 and 4–8h after ovulation. It was concluded that the optimal time of AI was 12–0h before ovulation using fresh semen and 4–0h before ovulation using frozen semen. In the Netherlands, Soede *et al.* (1995), also using transrectal ultrasonics to monitor ovulation, concluded that fertilization rate was optimal when insemination of fresh semen took place at between 0 and 24h before ovulation.

The importance of the interval from insemination to ovulation in gilts inseminated with frozen semen had previously been demonstrated by Schwarz (1993); for gilts inseminated >8, 4–8 and 0–4h before ovulation or 0–4h after ovulation, the fertilization rate was 0, 55, 88 and 50%, respectively. Suggested timings for fresh semen AI provided by Glossop (1991) are shown in Fig. 1.9.

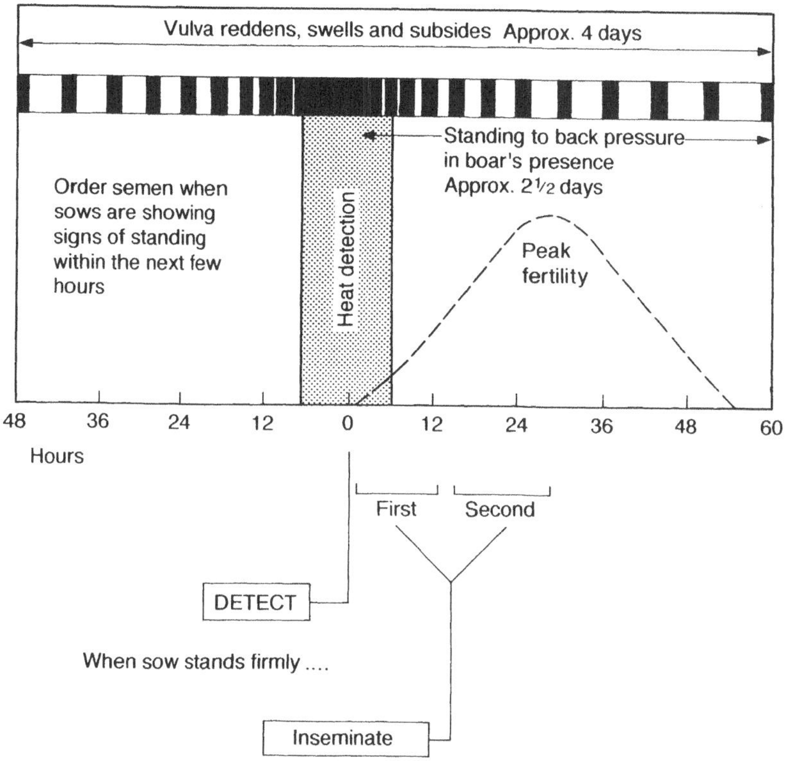

Fig. 1.9. Suggested timings to achieve optimal fertility with pig AI (from Glossop, 1991).

1.5.4. Seminal plasma and the timing of ovulation

It is now known that boar seminal plasma interacts with the female genital tract in such a way that the possibilities of fertilization are enhanced. Intrauterine infusion of seminal fluid at the start of the heat period was found to shorten the interval between the onset of oestrus and ovulation; this appeared to be an effective mechanism in reducing the variability in ovulation time in the animal (Everwand, 1990; Weitze *et al.*, 1990; Sudhoff, 1994). In other studies in Germany, Soares (1995) reported evidence that transcervical infusion of seminal plasma significantly increased the number of accessory sperm in the zonae pellucidae of oocytes after a subsequent insemination.

It is known that the mean number of accessory sperm in early embryos in the pig can be positively related to the percentage of normal embryos in the animal (Soede *et al.*, 1995). Studies reported by Weitze and Waberski (1996) examined components involved in the advancement of ovulation; seminal plasma was found to advance ovulation by 8–14 h. The authors concluded that seminal plasma advances ovulation by an additive effect of oestrogens and an unidentified peptide fraction via a local pathway between the uterus and ovary.

1.5.5. Climatic factors and AI results

Some countries have provided information on the effect of climatic factors on the reproductive performance of pigs bred by artificial insemination. In Germany, for example, Huhn *et al.* (1995) analysed results of 25,215 first inseminations carried out over a 3-year period on farms. The highest number of piglets born per 100 inseminations (892) occurred for maximum daily temperatures of 10–15°C during the preweaning period; for temperatures greater than 20°C, 25–30°C and greater than 30°C, the corresponding piglet figures were 830, 774 and 692, respectively. The number of hours of sunlight daily did not affect the reproductive traits, other than by its effect on temperature.

1.5.6. Semen diluents

The difficulties encountered in the cryopreservation of boar sperm in the 1950s (Polge, 1956) led to the examination of alternative procedures. The short-term preservation of boar semen in various diluents increased during the 1960s and this was a period marked by the development of several fresh semen diluents that were adapted for the commercial use of pig AI. The pig industry subsequently made extensive use of fresh semen stored for 2–4 days above 15°C (Watson, 1990). Semen diluents are designed to afford protection against temperature and changes in pH, increase semen volume, provide a source of nutrients for the sperm and inhibit bacterial growth. Sodium bicarbonate and sodium citrate are common buffers found in pig diluents. Antibiotics used in pig semen diluents include penicillin, streptomycin and gentamycin. Of the various diluents employed in commercial pig AI, one that was commonly used was based on ethylenediaminetetraacetate (EDTA) and usually referred to as the Kiev diluent (also Merck or Varohm diluent). The Kiev diluent proved popular because it was simple to prepare and boar semen could be used for up to 3 days without any appreciable loss in fertility.

A sperm dose of 2000–3000 million per insemination is commonly used, in a diluent volume of some 70 ml.

The commercial requirements for an effective fresh semen diluent have been described by Reed (1991). The author argued that, since no diluent can satisfy all commercial needs, it may be necessary to develop diluents for specific situations rather than for multipurpose use. The composition of some of the diluents used in fresh semen AI in the UK are given in Table 1.5. In China, Cheng (1988) reported on fertilization rates after AI with semen extended in various diluents and stored for 1–9 days; the percentage of fertilized oocytes 3 days after AI was 89.8–100% for Zorpva-extended semen and 55.8–100% for Kiev-extended semen. The effect of oxytocin added to the Kiev-extender was examined by Krajnak (1988) in former Czechoslovakia; differences due to oxytocin (5 IU 100 ml^{-1}) proved to be significant for the

Table 1.5. Composition of some boar semen diluents used in the UK (excluding antibiotics) (from Glossop, 1991). EDTA, ethylenediaminetetraacetate; PVA, polyvinylalcohol.

Diluent	Semen life (days)	Ingredients
Kiev	3	Glucose, sodium citrate, sodium bicarbonate, EDTA
BTS	4–5	Glucose, sodium citrate, sodium bicarbonate, EDTA, potassium chloride
Zorpva	5	Glucose, tri-sodium citrate, sodium bicarbonate, EDTA, TRIS, citric acid, cysteine, PVA
Reading	4–6	Glucose, tri-sodium citrate, sodium bicarbonate, EDTA, potassium chloride, TRIS, citric acid, cysteine, PVA, trehalose

sows inseminated in their study. Many comparisons of diluents for fresh semen pig AI have been made by investigators in the USA (Hunt *et al.*, 1994).

1.5.7. Long-term storage of semen

The development of semen diluents to prolong the life of stored pig sperm has been in progress since 1980; for pig AI purposes, extending the viability of boar sperm for longer than 3 days is regarded as important (Weitze, 1990, 1991). However, to minimize physical damage to sperm, diluted semen should be stored at a temperature above 15°C, but this impairs fertilizing ability. Although it is possible to store fresh semen for up to 4 days while maintaining the fertilizing ability of the sperm, after 5 or 6 days, fertilizing ability will decrease (Waberski *et al.*, 1994c); this decrease varies significantly among boars. It has been shown (Waberski *et al.*, 1994a) that the general assumption of constant fertilizing capacity of boar semen stored over a 3-day period in various extenders, including the Beltsville thawing solution (BTS) must be rejected; a decrease in fertilization rate was recorded between day 1 and day 2 semen. It was apparent that a decrease in fertilizing capacity due to the *in vitro* ageing of sperm was not prevented, even during the early days of storage.

Pooled semen

The use of pooled, rather than single sire semen, has been shown in some studies to give higher conception rates. In Germany, Bortolozzo (1992) reported that the percentage of oocytes fertilized was significantly higher for gilts inseminated with pooled semen than with semen from individual boars (83% vs. 74%). Similar findings have been reported in some sheep and cattle studies.

1.5.8. Methods for evaluating boar semen quality

Visual assessments

Traditionally, visual assessment of boar sperm motility has been the main means employed in the pig AI laboratory; the accuracy of such assessments is largely dependent on the technician's experience and skill. Although computer-aided semen analysis has been used in the research laboratory to objectively measure sperm motility, for the commercial scene there is the major drawback of cost. Some staining procedures may be of interest. In the USA, for example, Althouse and Hopkins (1995) have shown that sperm incubation with the fluorophore combination of calcein acetylmethyl ester (CAM) and ethidium homodimer (EH) provided an accurate method of objectively assessing individual boar sperm viablity by way of their distinctive fluorescent patterns in viable and non-viable sperm.

In vitro *assessments*

Although correlations of semen quality traits with fertility based on field trials are generally low, there is still a place for the *in vitro* assessment of pig semen quality (Woelders, 1991). *In vitro* assessment methods are based on sperm characters that are believed to be associated with fertilizing ability. The author cites examples such as membrane integrity, acrosome integrity, motility, stress sensitivity, rates of metabolic fluxes, activity of specific enzymes, metabolite concentrations, chromatin structure and characters related to sperm–oocyte interaction; such sperm assays usually provide data on a limited number of aspects of the fertilization process and in many instances a correlation to *in vivo* fertilizing potential may not be established. This is because fertility in field trials is measured by parameters, such as pregnancy rates, farrowing rates and litter sizes, which are likely to be influenced by many factors unrelated to semen quality. It is for such reasons that high insemination numbers are usually required for pig AI, adding to the cost of the technique.

Several reports have suggested that the zona-free hamster oocyte bioassay (Berger and Parker, 1989) or the *in vitro* penetration of *in vitro*-matured oocytes (Bacci *et al.*, 1991; Ivanova and Mollova, 1993) may be useful in the *in vitro* fertility assessment of fresh boar semen.

1.5.9. Frozen boar semen

After the successful use of glycerol in the freeze–thawing of bull sperm, Polge (1956) attempted to use the same approach to the freezing of boar semen. It was soon evident that boar sperm responded to cooling and freezing quite differently from bull sperm; either glycerol addition or cooling to temperatures below 15°C led to a marked reduction in sperm survival. It is now clear that differences in the sensitivity of mammalian sperm to cryopreservation make it almost impossible to extrapolate well-established techniques from one animal species to another. It was not until the 1970s that fertilization with frozen–

thawed boar semen after intraoviductal insemination was achieved (Polge *et al.*, 1970). This was followed by the success of several other groups using intracervical insemination (Crabo and Einarsson, 1971; Graham *et al.*, 1971; Pursel and Johnson, 1971).

By the mid-1970s, commercially applicable boar semen freezing procedures had been developed in various laboratories, including the Beltsville Pellet Freezing and Thawing Procedure (Pursel and Johnson, 1975) and the German Maxi Straw technique (Westendorf *et al.*, 1975). However, from several later reports (Johnson *et al.*, 1982; Johnson, 1985, 1989) it was clear that frozen semen had a much reduced fertility even when much larger sperm numbers were inseminated compared with fresh semen (50–60% farrowing rates and 8–9 piglets per litter). For these reasons, although freeze–thaw procedures are used commercially, this is only in very limited circumstances, such as for semen exports.

Surgical insemination of the oviducts was used in Illinois by Schoenbeck and Didion (1995a) in an effort to spare the use of valuable semen. Semen was frozen using the Beltsville Pelleting Method and placed in the oviduct 24h after the onset of oestrus. The total sperm injected varied from 0.5 to 1.0 million per oviduct and production averages from three farms were 29% farrowing rate and 5.9 piglets born alive per litter. The inability to increase fertility results to the levels usually expected with fresh semen (80–85% farrowing rate; 10–11 piglets per litter) has prevented frozen semen from being used commercially on any scale.

In the mid-1980s, Reed (1985) estimated that less than 50,000 doses of frozen boar semen were being produced and used on an annual basis and it is unlikely that the situation has changed greatly during the following decade. Although the use of AI with fresh semen is apparently gaining in popularity in the USA, the acceptance of frozen semen continues to be limited. Although the precise reasons for reduced reproductive performance after the use of frozen semen is unknown, they are believed to include (i) reduced lifespan and motility of frozen–thawed sperm, (ii) sensitivity of individual boar's sperm to cryopreservation, and (iii) the experience of the breeding personnel (Schoenbeck and Didion, 1995b). In Japan, Masuda *et al.* (1993) concluded that the main cause of a lower farrowing rate and litter size of pigs inseminated with frozen–thawed semen was the failure to fertilize ovulated oocytes rather than embryonic mortality.

According to Fonseca *et al.* (1995), the application of semen cryopreservation to boar reproduction may be made more efficient using semen from certain breeds. They showed, in studies with semen from five different breeds, that responses to cryopreservation may be correlated with *in vitro* fertilization, the cholesterol/phospholipid ratio and with the kinetics of *in vitro* decondensation of sperm nuclei.

It is known that the ability of the boar sperm head plasma membrane to exhibit the acrosome reaction may be affected by freeze–thawing and that this may be related to the decreased fertility found with frozen semen. There are those who believe that understanding the molecular damage to the plasma membranes in thawed boar sperm may facilitate development of effective cryoprotectants (Buhr, 1991). In other studies in Guelph, Pettitt and Buhr (1995) reported on the sensitivity of the head plasma membrane to components (detergents) in cryopreservation media. There is also evidence from studies in the UK of membrane changes resembling capacitation in boar sperm induced by cooling to low temperatures above freezing (Ashworth *et al.*, 1996); this may be a factor in explaining why cryopreserved or cooled boar sperm require shorter capacitation times or even bypass the need for capacitation entirely. In this regard, Renard *et al.* (1995) have described an unequivocal and rapid procedure for scoring acrosome-reacted boar sperm, using fluorescein-labelled GB 24, a mouse monoclonal antibody.

Research on various aspects of frozen boar semen has been reported from many countries in recent years, including Canada (Joyal *et al.*, 1986; Fiser, 1991), the USA (Hammitt and Martin, 1989a,b), France (Hamamah *et al.*, 1990), Sweden (Bwanga, 1991; Bwanga *et al.*, 1991a,b,c), Norway (Hofmo and Almlid, 1991), Germany (Weitze *et al.*, 1991), Austria (Berger and Fischerleitner, 1992), Japan (Bamba and Miyagawa, 1992), Bulgaria (Kovachev *et al.*, 1994) and the Netherlands (Fazeli *et al.*, 1995).

Changing sensitivity of boar sperm to cooling

When boar sperm are ejaculated, they are very prone to cooling damage. With the passage of time, they acquire resistance to cooling, a fact that is taken into account in the various protocols employed in cryopreservation. An interval of 2–4h must elapse before boar sperm are cooled below 15°C when they are being processed for freezing (Watson, 1990). Work in the UK, however, has demonstrated that cold resistance in boar sperm increases at room temperature for about 16 h after ejaculation (Tamuli and Watson, 1994). For this reason, it is suggested that during cryopreservation, some advantage may be gained by incubating sperm for a longer period than is commonly employed before cooling, in order to permit the full development of cryoresistance of the membranes.

1.5.10. The insemination procedure

The insemination technique has been described by Glossop (1991). The first step is the introduction of the insemination catheter through the lips of the vulva into the vagina (Fig. 1.10); the tip of the catheter may be lubricated before use and is directed upwards and forwards towards the spiral configuration of the cervix. Once in contact with the cervix, the catheter is twisted in an anticlockwise direction to lock it into the cervix. The semen container is then attached to the catheter and pressure gently applied over a period of several minutes to run the semen into the uterus.

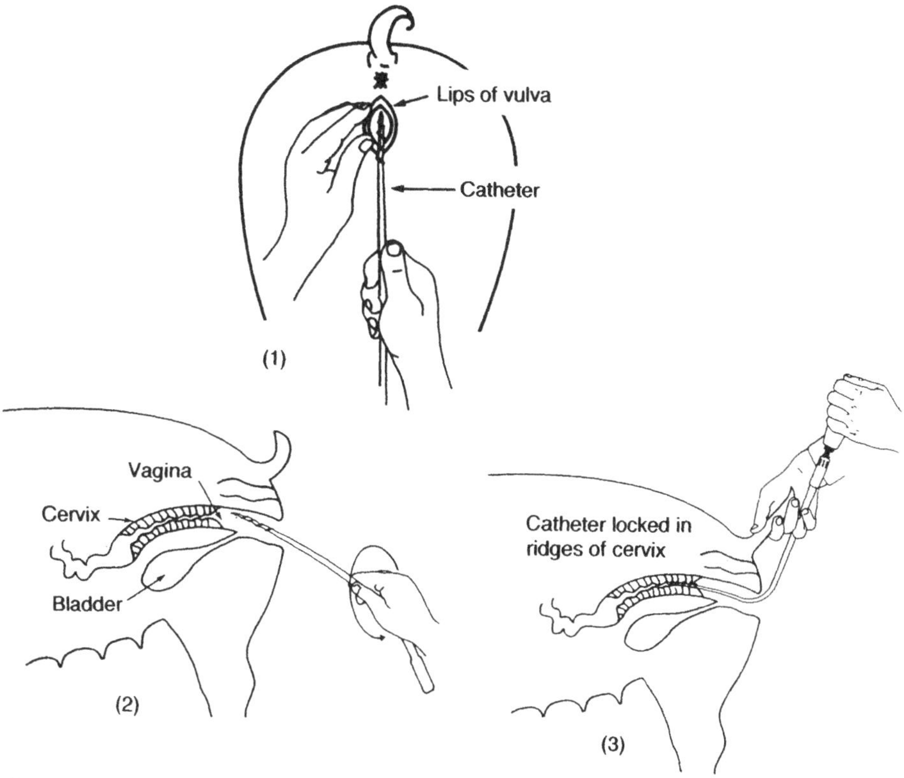

Fig. 1.10. Stages in the artificial insemination of the pig. (1) Introduction of catheter through lips of vulva; (2) twisting catheter in an anticlockwise direction to lock in place; (3) attaching insemination bottle and exerting gentle pressure on it (from Glossop, 1991).

1.5.11. Manipulation of immunological responses to enhance conception

There have been reports suggesting that the addition of specific and non-specific cellular antigens to semen at breeding may increase embryo survival up to the fifth week of pregnancy in the pig (Almlid, 1981). Other studies have indicated that presensitization of the uterus to sperm and seminal antigens several weeks prior to breeding may be of value in increasing litter size in gilts (Murray and Grifo, 1986). Although Lenge *et al.* (1986) in the Netherlands found that the addition of bovine leucocytes to boar semen neither influenced pregnancy rate nor the numbers of embryos surviving in early pregnancy, in Canada, Gooneratne and Thacker (1989) demonstrated that fertility in maiden gilts could be improved significantly by presensitization of the uterus to sperm and seminal antigens prior to breeding.

The major beneficial effect in this latter report was a significant increase in conception rate in gilts receiving an intrauterine infusion of attenuated semen during their first and second oestrus; the authors suggested that such an infusion could be a useful and practical method of improving the fertility of maiden gilts. In view of the fact that standard practice in many commercial pig units is to delay mating maiden gilts until the second or third oestrus because of an anticipated increase in ovulation rate, no production time need be lost by the application of the treatment. Further evidence of a positive effect came from the Czech Republic, where Cerovsky (1995) gave Large White gilts an infusion of dead sperm from Landrace boars at the oestrus prior to AI with live Landrace sperm; the author recorded a significantly higher conception rate (87.4 vs. 79.3% in controls) but no effect on litter size after such treatment.

1.5.12. *Relative costs and merits of using AI or natural service*

A theoretical study was made in France with herds of 56 to 420 sows by Salaun (1989) of the relative costs of using natural mating, AI with semen produced on the farm and AI with purchased semen. Costs declined with increasing herd size for all methods of reproduction. The use of farm-produced semen was less costly than natural mating for herds of more than 112 sows; the use of bought semen was less costly than natural mating for herds of 210 and 420 sows. A comparison of the cost of AI and natural mating carried out by Mani (1991) showed that although AI resulted in lower costs than natural service and brought benefits for selection and health, the conception rate of naturally bred sows was higher than that of inseminated pigs and they produced 0.2–0.5 more piglets per litter.

In the USA, Flowers and Alhusen (1992) dealt with reproductive performance and labour requirements associated with combinations of AI and natural service. When four or more sows and gilts were inseminated per day, the time required for inseminating an animal was significantly less than the time required for natural mating; gilts required significantly more time for mating than sows. The workers demonstrated that gilts and sows responded differently in reproductive performance to combinations of natural mating and AI.

1.6. Oestrus and Its Detection in the Pig

The effectiveness of oestrus detection is likely to depend on several factors, the most important of which would be the presence of a boar and the skill and experience of the stockperson. As noted by Winfield (1980), the only certain way of detecting heat is to ensure that the gilt or sow has daily contact with an active boar in a setting as free as possible from distractions; it is probably better to take sows to the boar when oestrus is expected rather than the boar to the sow (Fig. 1.11). This means that the boar pen must be situated reasonably close to the females, which, in itself, is likely to be a factor facilitating the full expression of oestrus.

Fig. 1.11. Identifying sexually receptive sows and gilts in the presence of the boar (from Hunter, 1982).

1.6.1. Detection in the absence of the boar

Without the boar, the stockperson's skill becomes all-important and it is necessary to know the behavioural and other characteristics which are associated with heat in the female pig. One behavioural characteristic which is often used as the basis of oestrus detection is the animal's response to the 'riding-test', which takes the form of the immobilization reflex in which the sow stands quite still, arches her back and pricks up her ears. According to Signoret (1972), at most 50–60% of oestrous pigs are likely to be detected in this way, in the absence of the boar. A further addition to the riding-test may be to manually simulate the nosing activity of the boar along the flanks of the sow, while applying pressure to the sow's back (Winfield, 1980).

1.6.2. Male pheromones

Already mentioned is the question of boar odour. Two of the steroids produced by the boar testis (3α-androstenol and 5α-androstenone) are believed to be concentrated in the submaxillary gland of the mature boar and secreted into the saliva, whereupon they act as pheromones to facilitate the induction of the mating stance in the oestrus sow. Pheromones are now available in aerosol form which can be used as an aid to obtaining the full immobilization reflex in the absence of a boar.

1.6.3. Measuring the conductivity of vaginal mucus

Evidence in the early 1980s showed that equipment which could be used to follow changes in the electrical conductivity of vaginal mucus, might be used to predict with some accuracy the optimum time to inseminate the pig with fresh semen (Zink and Diehl, 1984). In Canada, Harbison *et al.* (1987)

compared the fertility of sows following insemination with fresh or frozen-thawed semen when AI was at a fixed time after the onset of oestrus or when indicated to be optimal by a conductivity meter. They concluded that insemination according to a conductivity meter, although adequate for fresh semen, lacked the accuracy required when frozen semen is employed.

Other workers have reported experiences in the use of conductivity meters. In Taiwan, Huang *et al.* (1989) concluded that vaginal conductivity was a useful indicator of the optimum time of mating. In the USA, Ko *et al.* (1989) reported that a single mating at 24 or 36 h after a 15-point or higher vaginal mucosal conductivity increase (using an Ovu-Mate meter) resulted in a farrowing rate and litter size comparable to that of twice-mated control sows. In Canada, Charuest *et al.* (1990) recorded a lower conception rate and litter size in sows bred on the basis of vaginal conductivity than when natural service or AI was conducted on the basis of checking for oestrus with a boar. Studies on the electrical conductivity of the vaginal mucus have also been reported from the Czech Republic (Poschl and Rezac, 1992) and Italy (Tarocco and Gherpelli, 1992).

On a wider front, some authors have investigated the possibility of monitoring individual sows in group housing conditions for various functions. In the Netherlands, for example, Bressers (1993) looked at the automated detection of oestrus in group-housed sows, physiological activity associated with oestrus, the monitoring of individual sows from radiotelemetrically recorded ear-base temperature around farrowing and electronic sow feeding. In the USA, Blair *et al.* (1994b) evaluated the feasibility of delivering feed and detecting oestrous behaviour in computer-controlled equipment in a non-confinement environment; the behaviour of pigs, as measured by visits to a boar's pen, were recorded electronically. Results indicated a close relationship between visits to the boar pen and observed oestrus.

1.7. References

Aherne, F.X. and Williams, I.H. (1992) Nutrition for optimizing breeding herd performance. *Veterinary Clinics of North America, Food Animal Practice* 8(3), 589–608.

Almlid, T. (1981) Does enhanced antigenicity of semen increase the litter size in pigs? *Zeitschrift fur Tierzuchtg. Zuchtgbiol.* 98, 1–10.

Althouse, G.C. and Hopkins, S.M. (1995) Asssessment of boar sperm viability using a combination of two fluorophores. *Theriogenology* 43, 595–603.

Anon. (1993a) Thailand's AI increase. *Pig International* 23(9), 19.

Anon. (1993b) Pigs and goats. An increase in insemination. *Livestock from France Newsletter*, no. 4, 4.

Ashdown, R.R. and Hancock, J.L. (1974) Functional anatomy of male reproduction. In: Hafez, E.S.E. (ed.) *Reproduction in Farm Animals.* Lea and Febiger, Philadelphia, pp. 3–23.

Ashworth, C.J. (1990) Effect of pre-mating nutritional status and post-mating progesterone supplementation on early pregnancy in Large White gilts. *Animal Production* 50, 560.

Ashworth, C.J. and Atkinson, T. (1995) Relationships between foetal fluid insulin-like growth factor-one levels and foetal growth following reciprocal embryo transfer between Meishan and Landrace × Large White gilts. *Proceedings of the British Society of Animal Science (Winter Meeting)*, Paper no. 184.

Ashworth, C.J., Miller, S.J. and Flint, A.P.F. (1995) Conceptus and endometrial protein secretion in Meishan and Large White × Landrace gilts. *Journal of Reproduction and Fertility* Abstract Series no. 15, p. 56.

Ashworth, P.J.C., Harrison, R.A.P. and Watson, P.F. (1996) Membrane changes resembling capacitation in boar spermatozoa induced by cooling to low temperatures above freeezing. *Journal of the British Fertility Society* 1(1), 23A.

Aumaitre, A., Dogorn, J., Legault, C. and LeDenmat, M. (1976) Influence of farm management and breed types on sow's conception–weaning interval and productivity in France. *Livestock Production Science* 3, 75–83.

Bacci, M.L., Galeati, G., Mattioli, M. and Moretti, M. (1991) Evaluation of the fertilizing ability of boar semen by the zona pellucida penetration test. *Selezione Veterinaria* 32 (Suppl. 1), 309–314.

Bamba, K. and Miyagawa, N. (1992) Protective action of aromatic compounds against cold–shock injuries in boar spermatozoa. *Cryobiology* 29, 533–536.

Barnett, J.L. and Hemsworth, P.H. (1991) The effects of individual and group housing on sexual behaviour and pregnancy in pigs. *Animal Reproduction Science* 25, 265–273.

Bassett, J.M., Bray, C.J. and Sharpe, C.E. (1994) Melatonin administration and seasonal changes in plasma luteinizing hormone (LH) and prolactin (PRL) in a selected population of domestic sows kept outdoors without boars. *Journal of Reproduction and Fertility*, Abstract Series no. 14, pp. 7–8.

Benkov, M. (1989) Speculations on problems of reproduction on commercial pig complexes. *Selskostopanska Nauka* 27, 74–82.

Berger, B. and Fischerleitner, F. (1992) On deep freezing of boar semen: investigations on the effects of different straw volumes, methods of freezing and thawing extenders. *Reproduction in Domestic Animals* 27, 266–270.

Berger, T. and Parker, K. (1989) Modification of the zona-free hamster ova bioassay of boar sperm fertility and correlation with *in vivo* fertility. *Gamete Research* 22, 385–397.

Bichard, M. (1980) The breeding sow – today and tomorrow. *Report of the Agricultural Economics Unit, Exeter University* 209, 31–53.

Bichard, M. and David, P.J. (1986) Producing more pigs per sow per year – genetic contributions. *Journal of Animal Science* 63, 1275–1279.

Biggs, C., Tilton, J.E., Craigon, J., Foxcroft, G.R., Ashworth, C.J. and Hunter, M.G. (1993) Comparison of follicular heterogeneity and ovarian characteristics in Meishan and Large White hybrid pigs. *Journal of Reproduction and Fertility* 97, 263–269.

Blair, R. (1990) International swine production – capabilities and possibilities. *Outlook on Agriculture* 19, 263–268.

Blair, R.M., Coughlin, C.M., Minton, J.E. and Davis, D.L. (1994a) Peri-oestrous hormone profiles, embryonic survival and variation in embryonic development in gilts and primiparous sows. *Journal of Reproduction and Fertility* 101, 167–173.

Blair, R.M., Nichols, D.A. and Davis, D.L. (1994b) Electronic animal identification for controlling feed delivery and detecting estrus in sows in outside pens. *Journal of Animal Science* 72, 891–898.

Bortolozzo, F.P. (1992) Insemination of pigs with fresh, mixed semen, with reference to time of ovulation and age of semen. Thesis, Tierarztliche, Hochschule Hannover, 97 pp.

Bressers, H.P.M. (1993) Monitoring individual sows in group housing. Possibilities for automation. *Report No. B-396*, Schoonord Res. Institute, 135 pp.

Brzezinska-Slebodzinska, E., Slebodzinski, A.B., Pietras, B. and Wieczorek, G. (1995) Antoxidant effect of vitamin E and gluthathione on lipid peroxidation in boar semen plasma. *Biological Trace Element Research* 47, 69–74.

Buhr, M.M. (1991) Preservation of boar sperm alters membrane molecular dynamics. *Proceedings of the 2nd International Conference on Boar Semen Preservation* (Beltsville), pp. 81–93.

Bussiere, J. and Bariteau, F. (1992) Semen production in young Large White boars at an AI centre. In: *24es Journees de la Recherche Porcine en France (Paris)*, pp. 357–362.

Bwanga, C.O. (1991) Cryopreservation of boar semen. A literature review. *Acta Veterinaria Scandinavica* 32, 431–453.

Bwanga, C.O., Einarsson, S. and Rodriguez-Martinez, H. (1991a) Cryopreservation of boar semen. II. Effect of cooling rate and duration of freezing point plateau on boar semen frozen in mini- and maxi-straws and plastic bags. *Acta Veterinaria Scandinavica* 32, 455–461.

Bwanga, C.O., Hofmo, P.O., Grevle, I.S., Einarsson, S. and Rodriguez-Martinez, H. (1991b) *In vivo* fertilizing capacity of deep frozen boar semen packed in plastic bags and maxi-straws. *Journal of Veterinary Medicine, Series A* 38(4), 281–286.

Bwanga, C.O., Einarsson, S. and Rodriguez-Martinez, H. (1991c) Deep freezing of boar semen packaged in plastic bags and straws. *Reproduction in Domestic Animals* 26, 117–125.

Cameron, R.D.A. (1980) The effect of heat stress on reproductive efficiency in breeding pigs. *Veterinary Annual* 20, 259–264.

Campbell, K.H.S., McWhir, J., Ritchie, W.A. and Wilmut, I. (1996) Sheep cloned by nuclear transfer from a cultured cell line. *Nature* 380, 64–66.

Casida, L.E. (1935) Prepubertal development of the pig ovary and its relation to stimulation with gonadotrophic hormone. *Anatomical Record* 61, 389–396.

Cassar, G., Chapeau, C. and King, G.J. (1994) Effects of increased dietary energy after mating on developmental uniformity and survival of porcine conceptuses. *Journal of Animal Science* 72, 1320–1324.

Cerovsky, J. (1995) The effect of intrauterine infusion of dead spermatozoa after fertile insemination on litter size of gilts. *Zivocisna Vyroba* 40, 295–299.

Charuest, J.P., Dufour, J.J., Savoie, Y. and Richard, Y.M. (1990) Postweaning vaginal mucus conductivity and breeding performance at artificial insemination in swine. *Canadian Journal of Animal Science* 70, 451–457.

Cheng, W.T.K. (1988) Preservation of boar semen at 15°C. *Journal of Chinese Society of Veterinary Science* 14(4), 339–350.

Coffey, M.T. and Britt, J.H. (1993) Enhancement of sow reproductive performance by β-carotene or vitamin A. *Journal of Animal Science* 71, 1198–1202.

Crabo, B.G. (1985) Post-testicular sperm maturation and its importance to deep freezing of boar semen. *Proceedings of the 1st International Conference on Deep Freezing of Boar Semen* (Uppsala), pp. 25–27.

Crabo, B.G. (1991) Preservation of boar semen: a worldwide perspective. In: *Proceedings of the 2nd International Conference on Boar Semen Preservation* (Beltsville), pp. 3–12.

Crabo, B. and Einarsson, S. (1971) Fertility of deep frozen boar spermatozoa. *Acta Veterinaria Scandinavica* 12, 125–127.

Cunha, T.J. (1980) Action programs to advance swine production efficiency. *Journal of Animal Science* 51, 1429–1433.

D'Allaire, S. and Leman, A.D. (1990) Boar culling in swine breeding herds in Minnesota. *Canadian Veterinary Journal* 31, 581–583.

Davies, W. (1994) The porcine stress syndrome. *Animal Biotechnology* 5, 169–172.

Dial, G.D., Dial, O.K., Wilkinson, R.S. and Dziuk, P.J. (1984) Endocrine and ovulatory responses of the gilt to exogenous gonadotropins and estradiol during sexual maturation. *Biology of Reproduction* 30, 289–299.

Dirks, C. (1989) Fertility disorders in the boar. *Schweine-Zucht und Schweine-Mast* 37, 281–284.

Dourmad, J.Y., Etienne, M., Prunier, A. and Noblet, J. (1994) The effect of energy and protein intake of sows on their longevity: a review. *Livestock Production Science* 40, 87–97.

Downey, B.R. and Driancourt, M.A. (1994) Morphological and functional characteristics of preovulatory follicles in Large White and Meishan. *Journal of Animal Science* 72, 2099–2106.

Edwards, S. and Baxter, M. (1989) High-welfare systems for pigs. School of Agriculture, University of Aberdeen, *Annual Report*, pp. 11–16.

Edwards, S.A., Smith, W.J., Fordyce, C. and MacMenemy, F. (1994) An analysis of the causes of piglet mortality in a breeding herd kept outdoors. *Veterinary Record* 135, 324–327.

English, P. and Burgess, G. (1995) Its more than a job – some say it's an art. *Pig Farming* 43(12), 43.

English, P.R., Smith, W.J. and Maclean, A. (1984) *The Sow: Improving Her Efficiency*, 2nd edn, Farming Press, Ipswich.

Etienne, M. and Dourmad, J.Y. (1994) Effects of zearalenone or glucosinolates in the diet on reproduction in sows: a review. *Livestock Production Science* 40, 99–113.

Evans, L.E. and Ko, J.C.H. (1990) Electroejaculation and artificial insemination in Vietnamese potbellied miniature pigs. *Journal of the American Veterinary Medical Association* 197, 1366–1367.

Everwand, A. (1990) Effect of different sperm-free inseminates on the time of ovulation, in relation to the time of stimulation in the sow. Thesis, Tierarztliche Hochschule Hannover, 79 pp.

Faillace, L.S. and Hunter, M.G. (1994) Follicle development and oocyte maturation during the immediate preovulatory period in Meishan and white hybrid gilts. *Journal of Reproduction and Fertility* 101, 571–576.

Fazeli, A.R., Holt, C., Steenweg, W., Bevers, M.M., Holt, W.V. and Colenbrander, B. (1995) Development of a sperm hemizona binding assay for boar semen. *Theriogenology* 44, 17–27.

Filatov, A.I. and Rybalko, V.P. (1990) The organization of pig breeding in China. *Zootekhniya*, no. 4, 75–79.

Fiser, P.S. (1991) Interactions of cooling velocity, warming velocity and glycerol concentration on the survival of frozen–thawed boar sperm. *Proceedings of the 2nd International Conference on Boar Semen Preservation* (Beltsville), pp. 123–137.

Flowers, W.L. (1992) Artificial insemination in swine. *Agri-Practice* 13, 36–40.

Flowers, W.L. (1995) Characterization of the implementation and use of artificial insemination (AI) in the N.C. swine industry. *Journal of Animal Science* 73 (Suppl. 1), 6.

Flowers, W.L. and Alhusen, H.D. (1992) Reproductive performance and estimates of labor requirements associated with combinations of artificial insemination and natural service in swine. *Journal of Animal Science* 70, 615–621.

Fonseca, T., Doculomb, Y., Campos, C., Gomez, J.L., Sierra, T., Betancourt, M. and Rosado, A. (1995) Differential susceptibility to cryopreservation of spermatozoa obtained from different boar breeds: role of cholestrol/phospholipids (Chol/Phos) ratio on fertilization and nuclear decondensation. *Biology of Reproduction* 52 (Suppl. 1), p. 122.

Ford, S.P. (1995) Specific effects of the Meishan uterus on decreasing embryonic cell size and oestradiol-17B secretory ability. *Journal of Reproduction and Fertility* Abstract Series No. 15, p. 36.

Fowler, V.R. (1980) The future of the pig as a meat animal. *Proceedings of the Nutrition Society* 39, 151–159.

Fuentes, P.A.R., Serrano, G.L.De., Manzo, M.R.De., Regueiro, C. and Valle, A. (1993) Effect of season on semen traits of boars in the tropics. *Zootecnia Tropical* 10(1), 51–64.

Gerfen, R.W., White, B.R. and Wheeler, M.B. (1992) Comparison of semen characteristics between Fegjing, Meishan and Yorkshire boars. *Biology of Reproduction* 46 (Suppl. 1), p. 146.

Gerrits, R.J., Blosser, T.H., Purchase, H.G., Gerill, C.E. and Warwick, E.J. (1979) Economics of improving reproductive efficiency in farm animals. In: Hawk, H. (ed.) *Animal Reproduction Symposium* no.3, pp. 413–421.

Glossop, C.E. (1991) Pig artificial insemination re-assessed. *In Practice* 13(5), 191–195.

Glover, T. and Nicander, L. (1971) Some aspects of structure and function in the mammalian epididymis. *Journal of Reproduction and Fertility* Suppl. 13, 39–50.

Gooneratne, A.D. and Thacker, P.A. (1989) Effect of intrauterine infusion of sperm antigens on gilt fertility. *Theriogenology* 31, 1221–1226.

Graham, E.F., Rahamannan, A.H.J., Schinehel, M.K.L., Waki-Laorial, N. and Bower, R.E. (1971) Preliminary report on procedure and rationale for freezing boar semen. *AI Digest* 19, 12–14.

Gronbach, G. (1995) Artificial insemination of pigs in Germany. *Kleinviehzuchter* 43, 725–728.

Haley, C.S. (1991) Considerations in the development of future pig breeding programs. *Asian–Australian Journal of Animal Sciences* 4(4), 305–308.

Haley, C.S. (1996) Markers, maps and the detection of trait loci. *Proceedings of the British Society of Animal Science* (*Winter Meeting*), Paper no. 50.

Haley, C.S., Lee, G.J. and Ritchie, M. (1995) Comparative reproductive performance in Meishan and Large White pigs and their crosses. *Animal Science* 60, 259–267.

Hamamah, S., Royere, D., Nicolle, J.C., Paquignon, M. and Lansac, J. (1990) Effects of freeze–thawing on the spermatozoon nucleus: a comparative chromatin cytophotometric study in the porcine and human species. *Reproduction Nutrition and Development* 30, 59–64.

Hammitt, D.G. and Martin, P.A. (1989a) Fertility of frozen–thawed porcine semen following controlled-rate freezing in straws. *Theriogenology* 32, 359–368.

Hammitt, D.G. and Martin, P.A. (1989b) Correlations among assays of porcine semen quality following cryopreservation. *Theriogenology* 32, 369–384.

Harayama, H. and Kato, S. (1993) Characteristics of the ejaculates from Meishan boars at various ages. *Animal Science and Technology* 64, 333–339.

Harbison, S., Kirkwood, R.N., Aherne, F.X. and Sather, A.P. (1987) Reproductive performance of sows inseminated with fresh or thawed semen: influence of sows after insemination at a fixed time or as indicated by changes in vaginal mucus conductivity. *Canadian Journal of Animal Science* 67, 865–867.

Hemsworth, P.H. and Barnett, J.L. (1990) Behavioural responses affecting gilt and sow reproduction. *Journal of Reproduction and Fertility*, Suppl. 40, 343–354.

Hemsworth, P.H. and Winfield, C.G. (1986) Sexual behaviour problems in the male and female pig. In: Morrow, D.A. (ed.) *Current Therapy in Theriogenology*. W.B. Saunders, Philadelphia, pp. 915–918.

Hemsworth, P.H., Brand, A. and Willems, P. (1981) The behavioural response of sows to the presence of human beings and its relationship to productivity. *Livestock Production Science* 8, 67–74.

Hemsworth, P.H., Barnett, J.L., Coleman, G.J. and Hansen, C. (1989) A study of the relationships between the attidudinal and behavioural profiles of stockpersons and the level of fear of human and reproductive performance of commercial pigs. *Applied Animal Behaviour Science* 23, 301–314.

Hoagland, T.A. and Diekman, M.A. (1982) Influence of supplemental lighting during increasing daylength on libido and reproductive hormones in prepubertal boars on growth and puberty in boars. *Journal of Animal Science* 55, 1483–1489.

Hofmo, P.O. and Almlid, T. (1991) Recent developments in freezing of boar semen with special emphasis on cryoprotectants. In: *Proceedings of the 2nd International Conference Boar Semen Preservation* (Beltsville), pp. 111–121.

Honeyman, M.S. (1994) Sustainability issues of U.S. swine production. *Journal of Animal Science* 72 (Suppl. 1)/*Journal of Dairy Science* 77 (Suppl. 1), p. 149.

Huang, Y.T., Chung, C.F. and Lo, L.L. (1989) Using vaginal conductivity to determine the optimum breeding time in pigs. *Journal of the Chinese Society for Animal Science* 18, 67–73.

Hughes, P. and Varley, M. (1980) *Reproduction in the Pig*. Butterworths, London.

Huhn, U., Henze, A. and Jurk, R. (1995) Influence of selected ambient climatic factors on performance of inseminated primparous sows under large farm conditions. *Zuchttungskunde* 67, 305–316.

Hunt, T.H., N'Diaye, M.R. and Crabo, B.G. (1994) Comparison of extenders for liquid boar semen artificial insemination. *Journal of Animal Science* 72 (Suppl. 1)/*Journal of Dairy Science* 77 (Suppl. 1), p. 280.

Hunter, R.H.F. (1991) Fertilization in the pig and horse. In: Dunbar, B.S. and O'Rand, M.G. (eds) *A Comparative Overview of Mammalian Fertilization*. Plenum Press, New York, pp. 329–349.

Hunter, R.H.F. and Dziuk, P.J. (1968) Sperm penetration of pig eggs in relation to the timing of ovulation and insemination. *Journal of Reproduction and Fertility* 15, 199–208.

Hunter, M.G., Biggs, C., Foxcroft, G.R., McNeilly, A.S. and Tilton, J.E. (1993) Comparisons of endocrinology and behavioural events during the periovulatory period in Meishan and Large-White hybrid gilts. *Journal of Reproduction and Fertility* 97, 475–480.

Hunter, M.G., Biggs, C., Pickard, A.R. and Faillace, L.S. (1994a) Differences in follicular aromatase activity between Meishan and Large-White hybrid gilts. *Journal of Reproduction and Fertility* 101, 139–144.

Hunter, M.G., Faillace, L.S. and Picton, H.M. (1994b) Intrauterine and peripheral steroid concentrations and conceptus development in Meishan and Large White hybrid gilts. *Reproduction, Fertility and Development* 6, 783–789.

Hunter, M.G., Picton, H.M., Biggs, C., Mann, G.E., Faillace, L.S., McNeilly, A.S. and Foxcroft, G.R. (1995) Periovulatory hormone profiles in high ovulating Meishan sows and in Large White hybrid sows. *Journal of Reproduction and Fertility*, Abstract Series no.15, Abstract 188.

Hurtgen, J.P. and Leman, A.D. (1981) Effect of parity and season of farrowing on the subsequent farrowing interval of sows. *Veterinary Record* 108, 199–123.

Hurtgen, J.P., Leman, A.D. and Crabo, B. (1980) Seasonal influence of estrous activity in sows and gilts. *Journal of the American Veterinary Medical Association* 170, 199–223.

Hyttel, P., Nissen, A.K., Lehn-Jensen, H. and Greve, T. (1994) Oocyte maturation and conception in pigs. Cell biology and practical aspects. *Dansk Veterinaertidsskrift* 77, 709–713.

Ivanova, M. and Mollova, M. (1993) Zona-penetration in vitro test for evaluating boar sperm fertility. *Theriogenology* 40, 297–410.

Jindal, R. and Foxcroft, G.R. (1995) Progesterone: a mediator of nutritionally induced effects on embryo survival in gilts. *Journal of Animal Science* 73 (Suppl. 1), p. 306.

Johnson, L.A. (1985) Fertility results using frozen boar spermatozoa: 1970–1985. In: Johnson, L.A. and Larsson, K. (eds) *Deep Freezing of Boar Semen*. Swedish University of Agricultural Science, Uppsala, pp. 199–222.

Johnson, L.A. (1989) Current status of technological research and field application of frozen and fresh boar semen worldwide. *Japanese Journal of Swine Science* 26, 37–42.

Johnson, L.A., Aalbers, J.G. and Arts, J.A.M. (1982) Use of boar spermatozoa for artificial insemination. I. Fertilizing capacity of fresh and frozen spermatozoa in sows on 36 farms. *Journal of Animal Science* 52, 1130–1136.

Johnson, R.K., Zimmerman, D.W. and Kittok, R.J. (1984) Selection for components of reproduction in swine. *Journal of Animal Science* 11, 541–545.

Joyal, S.M., Kennedy, B.W. and Wilkins, J.N. (1986) Boar, breed and environmental effects on motility of frozen–thawed spermatozoa. *Canadian Journal of Animal Science* 66, 663–668.

Kaminski, M.A., Conley, A.J. and Ford, S.P. (1995) Developmental expression of cytochrome P450 17alpha-hydroxylase (P450c17) in day 10.5-14 Meishan and Yorkshire conceptuses. *Biology of Reproduction* 52 (Suppl. 1), p. 181.

Kano, K., Miyano, T. and Kato, S. (1994) Ovarian response to exogenous gonadotropins and in vitro fertilization of follicular oocytes in prepubertal Chinese Jinhua pigs. *Journal of Reproduction and Development* 40, 91–97.

Kemp, B. (1991) Nutritional strategy for optimal semen production in boars. *Pig News and Information* 12, 555–558.

Kemp, B. and Hartog, L.A. Den (1990) Influence of energy and protein intake on the fertility of the breeding boar. *Ubersichten zur Tierernahrung* 18, 77–90.

Kemp, B., Bakker, G.C.M., Hartog, L.A. Den and Verstegen, M.W.A. (1991) The effect of semen collection frequency and food intake on semen production in breeding boars. *Animal Production* 52, 355–360.

Kerr, J.C. (1995) Improving reproductive performance in pigs. *Proceedings of the British Society of Animal Science* (Winter Meeting) Paper no. 13.

Kerr, J.C. and Cameron, N.D. (1994) Reproductive performance of pigs selected for components of efficient lean growth. *Proceedings of the British Society of Animal Production* (Winter Meeting) Paper no.14.

Kerr, J.C. and Cameron, N.D. (1995) Genetic and phenotypic relationships between performance test and reproduction traits in pigs selected for components of efficient lean growth rate. *Proceedings of the British Society for Animal Science* (Winter Meeting) Paper no. 9.

King, J.W.B. (1970) Organization and practice of pig improvement in European countries. *Animal Breeding Abstracts* 38, 523–536.

King, R.H. (1993) The influence of nutrition on non-productive days in the sow. In: Farrell, D.J. (ed.) *Recent Advances in Animal Nutrition in Australia*, University New England, Armidale, pp. 233–238.

Kirkwood, R.N. and Thacker, P.A. (1988) Nutritional factors affecting embryo survival in pigs (results and speculation). *Pig News and Information* 9, 15–21.

Ko, J.C.H., Evans, L.E. and Hopkins, S.M. (1989) Vaginal conductivity as an indicator for optimum breeding time in the sow after weaning. *Theriogenology* 32, 961–968.

Ko, J.C.H., Evans, L.E. and Althouse, G.C. (1993) Toxicity effects of latex gloves on boar spermatozoa. *Swine Health and Production* 1(3), 24–26.

Kovachev, K.D., Zagorski, D.I., Ivanova, M.G. and Bobadov, N.D. (1994) Cryogenic damage to boar spermatozoa frozen in pellets and in tubes. *Theriogenology* 42, 1369–1379.

Krajnak, P. (1988) The effect of oxytocin added to insemination doses on fertility of pigs. *Zivocisna Vyroba* 33, 845–850.

Kuhlers, D.L., Jungst, S.B., Marple, D.N. and Rahe, C.H. (1985) The effect of pen density during rearing on subsequent reproductive performance in gilts. *Journal of Animal Science* 61, 1066–1069.

Kupferschmied, H.U. and Luginbuhl, J. (1990) 25 years of pig AI. Introduction into Switzerland and subsequent development. *Kleinviehzuchter* 38, 589–592.

Kyle, R. (1995) Wild boar in Britain: a new farming enterprise. *State Veterinary Journal* 5, 10–12.

Lean, I.J. (1988) Pigs. In: *Management and Welfare of Farm Animals*, 3rd edn. Universities Federation for Animal Welfare Handbook, Baillière Tindall, London, pp. 143–173.

Lee, G.J. and Haley, C.S. (1995) Comparative farrowing to weaning performance in Meishan and Large White pigs and their crosses. *Animal Science* 60, 269–280.

Lee, G.J., Ritchie, M., Thomson, M., MacDonald, A.A., Blasco, A., Santacreu, M.A., Argente, M.J. and Haley, C.S. (1995) Uterine capacity and prenatal survival in Meishan and Large White pigs. *Animal Science* 60, 471–479.

Leman, A.D. and Rodeffer, H.E. (1976) Boar management. *Veterinary Record* 98, 457–459.

Lenge, T. Van der, Donker, R. and Hazeleger, W. (1986) Embryonic survival in the pig after insemination with antigen-enriched semen. *Theriogenology* 26, 391–396.

Levis, D.G., Ford, J.J. and Christenson, R.K. (1989) *Boar sexual behaviour and evaluation technique*. Swine Research Progress Report, US Meat Animal Research Center, no. 3, pp. 39–41.

Liptrap, R.M. and Viveiros, M.M. (1993) Stress and changes in follicle development in sows. *Revista Brasileira de Reproducao Animal, Supplemento* no. 4, 131–142.

Liptrap, R.M., Castro, L. and Basrur, P.K. (1993) Stress and embryo mortality in sows. *Revista Brasileira de Reproducao Animal, Supplemento* no. 4, 108–131.

Liu, T., Ge, Y., Cheng, Q., Li, Y., Hao, G. and Li, Y. (1994) Advances in research on Chinese pigs in the USA. *Jiangsu Journal of Agricultural Sciences* 10, 37–42.

Louis, G.F., Lewis, A.J., Ermer, P.M., Stroup, W.W., Weldon, W.C. and Miller, P.S. (1992) Effect of energy and protein intakes on boar libido and semen characteristics. *Journal of Animal Science* 70 (Suppl. 1), 68.

Love, R.J. and Wilson, M.R. (1990) Fertility of sows mated on different days of the week. *Australian Veterinary Journal* 67(1), 1–3.

Love, R.J., Peacock, A.J. and Evans, G. (1990) Premature farrowings caused by feeding cottonseed meal. *Australian Veterinary Journal* 67, 223–226.

Love, R.J., Evans, G. and Klupiec, C. (1993) Seasonal effects on fertility in gilts and sows. *Journal of Reproduction and Fertility* Suppl. 48, 191–206.

Lynch, P.B., Ryan, D.P., Boland, M.P. and Roche, J.F. (1995) The effect of dietary fat supplementation on follicular development, ovulation rates and embryo survival in the sow. *Proceedings of the Irish Grassland and Animal Production Association* (21st Annual Meeting), pp. 111–112.

Macdonald, G.J., Wise, T., Sluss, P.M., Lunstra, D.D. and Ford, J.J. (1995) Differences in pituitary gonadotrophins from male pigs representing divergent breeds. *Biology of Reproduction* 52 (Suppl. 1), 77.

Mani, R. (1990) 25 years of pig AI. 3. A look abroad. *Kleinviehzuchter* 38, 593–595.

Mani, R. (1991a) Delegate meeting of the Swiss Association of Artificial Insemination. *Kleinviehzuchter* 39(4), 13.

Mani, R. (1991b) Explosive development of pig AI in France. *Kleinviehzuchter* 39, 1061–1062.

Mani, R. (1991c) A comparison of the cost of AI and natural mating. *KB-Mitteilungen* 29, 35–37.

Martinat-Botte, F., Bariteau, F., Badouard, B. and Terqui, M. (1985) Control of pig reproduction in a breeding programme. *Journal of Reproduction and Fertility, Suppl.* 33, 211–228.

Masuda, H., Oguri, N., Yonemura, I., Izumiya, K., Endo, K., Hoshina, K., Tomizuka, T. and Kojima, T. (1993) Factors influencing conception rate and litter size in gilts inseminated with frozen boar semen. *Japanese Journal of Swine Science* 30, 220–225.

Mauget, R. (1985) Seasonality of reproduction in the wild boar. In: Cole, D.J.A. and Foxcroft, G.R. (eds) *Control of Pig Reproduction.* Butterworths, London, pp. 509–526.

Maxwell, C.V., Combs, G.E., Knabe, D.A., Kornegay, E.T. and Nolan, P.R. (1994) Effect of dietary chlortetracycline during breeding and/or farrowing and lactation on reproductive performance of sows: a cooperative study. *Journal of Animal Science* 72, 3169–3176.

McGlone, J.J., Shaw, M. and Wilson, E. (1994) Outdoor versus indoor intensive pork production: pig production. *Journal of Animal Science* 72 (Suppl. 1)/*Journal of Dairy Science* 77 (Suppl. 1), p. 323.

McIntosh, G.B. (1989) The role of AI in the pig industry. *Queensland Agricultural Journal* 115, 142–146.

Meuwissen, T.H.E. and Van Arendonk, J.A.M. (1992) Potential improvements in rate of genetic gain from marker-assisted selection in dairy cattle breeding schemes. *Journal of Dairy Science* 75, 1651–1659.

Miller, A.T., Picton, H.M., Craigon, J. and Hunter, M.G. (1995) The dynamics of follicle development and oestrogenic activity in high ovulating Meishan sows and in Large White hybrid contemporaries. *Journal of Reproduction and Fertility*, Abstract Series no. 15, p. 13.

Miyano, T., Akamatsu, J., Kato, S., Nanjo, I. and Kanda, S. (1990) Ovarian development in Meishan pigs. *Theriogenology* 33, 769–775.

Mount, L.E. (1968) Adaptation of swine. In: Hafez, E.S.E. (ed.) *Adaptation of Domestic Animals.* Lea and Febiger, Philadelphia, pp. 277–291.

Mudra, K., Peter, W., Wegner, B., Potsch, H. and Traber, H. (1990) Investigations on the effects of a constant light regime on the performance of AI boars. *Archiv fur Tierzucht* 33, 65–75.

Murray, F.A. and Grifo, A.P. Jr (1986) Intrauterine infusion of killed semen to increase litter-size in gilts. *Journal of Animal Science* 62, 187–190.

Oldigs, B., Link, M., Baulain, U. and Kallweit, E. (1995) Investigation on outdoor piglet production in respect to animal welfare. First communication: evaluation by performance, technopathies and physiological reactions. *Zuchtungskunde* 67, 230–244.

Omeke, B.C.O. (1989) A comparison of seasonal effects on fertility of Landrace and Large White sows in subhumid tropical environment. *British Veterinary Journal* 145, 462–466.

Pattison, H.D. (1980) Patterns of sow culling. *Pig News and Information* 1(3), 215–218.

Pedersen, L.J. and Jensen, K.H. (1989) The influence of housing-systems for pregnant sows on the reproductive behaviour at oestrus. *Acta Agriculturae Scandinavica* 39, 331–343.

Pepper, T.A. and Taylor, D.J. (1981) Farrowing intervals of sows. *Veterinary Record* 108, 195.

Petchey, A.M. and Hunt, K.A. (1990) The boar: size and space requirements. *Farm Building Progress*, No. 99, 17–20.

Pettigrew, J.E. and Tokach, M.D. (1991) Nutrition and female reproduction. *Pig News and Information* 12(4), 559–562.

Pettitt, M.J. and Buhr, M.M. (1995) Membrane domains in sperm are differentially sensitive to semen cryopreservation media. *Biology of Reproduction* 52 (Suppl. 1), 57.

Pickard, A.R. and Ashworth, C.J. (1995) Oestradiol-17 β secretion by Chinese Meishan and Large White × Landrace conceptuses. *Journal of Reproduction and Fertility*, Abstract Series no. 15, p. 18.

Polge, C. (1956) Artificial insemination in pigs. *Veterinary Record* 68, 62–76.

Polge, C., Salamon, S. and Wilmut, I. (1970) Fertilizing capacity of boar semen following surgical insemination. *Veterinary Record* 87, 424–428.

Poschl, M. and Rezac, P. (1992) Vaginal impedance as a means of predicting ovulation in sows. *Sbornik-Jihoceska Univerzita, Zemedelska Fakulta, Ceske Budejovice, Zootechnicka Rada* 9, 80–81.

Potter, R. (1995) Investigation into boar infertility (1) Clinical examination. *Pig Journal* 35, 28–233.

Prunier, A., Chopineau, M., Mountier, A.M. and Mormede, P. (1993) Patterns of plasma LH, FSH, oestradiol and corticosteroids from birth to the first oestrous cycle in Meishan gilts. *Journal of Reproduction and Fertility* 98, 313–319.

Pursel, V.G. and Johson, L.A. (1971) *Procedure for the preservation of boar spermatozoa by freezing*. United States Department of Agriculture, Animal Research Service, 44, 227.

Pursel, V.G. and Johnson, L.A. (1975) Freezing of boar spermatozoa. Fertilizing capacity with concentrated semen and a new thawing procedure. *Journal of Animal Science* 40, 99–102.

Rasbech, N.O. (1969) A review of the causes of reproductive failure in swine. *British Veterinary Journal* 12, 599–616.

Reed, H.C.B. (1985) Current use of frozen boar semen – future need of frozen boar semen. *Proceedings of the 1st International Conference on Deep Freezing of Boar Semen (Uppsala)*, pp. 225–237.

Reed, H.C.B. (1991) Commercial requirements for an effective fresh semen diluent.

Proceedings of the 2nd International Conference Boar Semen Preservation (Beltsville), pp. 255–270.

Renard, P., Drenou, B., Griveau, J.F. and Lannou, D.Le (1995) Assessment of acrosome-reacted boar spermatozoa using monoclonal antibody GB24 and propidium iodide. *Theriogenology* 43, 927–938.

Rickard Bell, C.V., English, P.R., Dunne, J.H., MacPherson, O., Roden, J.A., Davidson, F.M. and Bamlett, R. (1993) Sources of inefficiency at mating in pigs – an appraisal. *Proceedings of the British Society for Animal Production* (Winter Meeting), Paper no. 132.

Rothschild, M.F. (1990) The role of biology in future pig breeding programs. *Proceedings of the 4th World Congress on Genetics Applied to Livestock Production* (Edinburgh), 15, 415–424.

Rothschild, M.F., Jacobson, D.A., Vase, D.A., Tuggle, C.K., Short, T.H., Sasaki, S., Eckardt, G.R. and McLaren, D.G. (1994) A major gene for litter size in pigs. In: *5th World Congress on Genetics Applied to Livestock Production* 21, 225–228.

Russo, V., Davoli, R., Tagliavini, J., Dall'Olio, S., Bigi, D., Costosi, E., Coscelli, M.B. and Fontanesi, L. (1993) Identification of the halothane sensitivity genotype of pigs at the DNA level, using the PCR. *Zootecnica e Nutrizione Animale* 19, 89–92.

Saito, O. and Sugiyama, K. (1990) The innovations of production technology and the conditions for successful management of large scale farms in the US hog industry. *Japanese Journal of Swine Science* 27, 194–201.

Salaun, Y. (1989) Comparison of the cost of artificial insemination, using bought semen or semen produced on the farm, with that of natural mating. *Techni-Porc* 12(6), 29–36.

Schoenbeck, R.A. and Didion, B.A. (1995a) Fertility of frozen boar sperm when deposited surgically in the oviduct. *Theriogenology* 43, 318.

Schoenbeck, R.A. and Didion, B.A. (1955b) Fertility of frozen boar sperm used for artificial insemination in a commercial setting. *Journal of Animal Science* 73 (Suppl. 1), 251.

Schook, L.B., Paszek, A.A., Louis, C., Murtaugh, M., Beattie, C.W., Rohrer, G.A., Alexander, L.J. and Wheeler, M.B. (1994) Mapping the swine genome: take home lessons from mouse and man. *Animal Biotechnology* 5, 129–134.

Schwarting, G. and Kleiner, B. (1995) Pig production in the year 2000. *Praktische Tierarzt* 76, 604–616.

Schwarz, M. (1993) Importance of the interval from insemination to ovulation in gilts inseminated with frozen semen. Thesis, Tierarztliche Hochschule Hannover, 69 pp.

Signoret, J.P. (1972) The mating behaviour of the sow. In: Cole, D.J.A. (ed.) *Pig Production*. Butterworths, London.

Signoret, J.P. (1989) Differentiation of sexual behaviour in young pigs. *Journees de la Recherche Porcine en France* 21, 129–132.

Signoret, J.P., Baldwin, B.A., Fraser, D. and Hafez, E.S.E. (1975) The behaviour of swine. In: Hafez, E.S.E. (ed.) *The Behaviour of Domestic Animals*. Baillière Tindall, London, pp. 294–329.

Sinclair, A.G., Edwards, S.A., Hoste, S. and McCartney, A. (1996) Evaluation of maternal and piglet influences on performance of Meishan synthetic and European White breeds of pig during lactation. *Proceedings of the British Society of Animal Science* (Winter Meeting), Paper no.19.

Soares, J.A.G. (1995) Fertilization results in pigs after transcervical use of seminal plasma immediately before insemination. Thesis, Tierarztliche Hochschule Hannover, Germany, 86 pp.

Soede, N.M. (1993) Boar stimuli around insemination affect reproductive processess in pigs: a review. *Animal Reproduction Science* 32, 107–125.

Soede, N.M., Wetzels, C.C.H., Zondag, W., Koning, M.A.I de. and Kemp, B. (1995) Effects of time of insemination relative to ovulation, as determined by ultrasonography, on fertilization rate and accessory sperm count in sows. *Journal of Reproduction and Fertility* 104, 99–106.

Solter, D. (1996) Lambing by nuclear transfer. *Nature* 380, 24–25.

Southwood, O.I., Hoste, S., Short, T.H., Mileham, A.J. and Cuthbert-Heavens, D. (1996) Evaluation of genetic markers for litter size in Meishan synthetic and Large White pigs. *Proceedings of the British Society of Animal Science* (Winter Meeting), Paper no.18.

Stolba, A. and Wood-Gush, D.G.M. (1984) The identification of behavioural key features and their incorporation into a housing design for pigs. *Annales des Recherches Veterinaires* 15, 287–298.

Stork, M.G. (1979) Seasonal reproductive inefficiency in large pig breeding units in Britain. *Veterinary Record* 104, 49–52.

Sudhoff, H. (1994) Ovulation induction by seminal plasma components in gilts. Thesis, Tierarztliche Hochschule Hannover, Germany, 79 pp.

Symonds, H.W. and Prime, G. (1989) The influence of volume of food intake by gilts on blood flow in the portal vein and clearance of progesterone from plasma. *Animal Production* 48, 620–621.

Tamuli, M.K. and Watson, P.F. (1994) Cold resistance of live boar spermatozoa during incubation after ejaculation. *Veterinary Record* 135, 160–162.

Tarocco, C. and Gherpelli, M. (1992) Studies on the electrical conductivity of the vagina in sows to determine the optimum time for insemination. *Selezione Veterinaria* 33, 357–368.

Tarn, C.Y., Rosenkrans, C.F., Brown, A.H., Rorie, R. and Kirby, J.D. (1995) Growth rate and testis size of boars treated with 6-propyl-2-thiouracil post-weaning. *Journal of Animal Science* 73 (Suppl. 1), 8.

Terqui, M., Bazer, F.W. and Martinat-Botte, F. (1992) Hyperprolificacy in Meishan sows. Mechanisms involved in early gestation. In: *24es Journees de la Recherche Porcine en France, Paris*, pp. 351–356.

Tsuma, V.T., Einarsson, S., Madej, A. and Kindahl, H. (1994) Effects of fasting sows during early pregnancy: patterns and embryonic survival. *Journal of Reproduction and Fertility*, Abstract Series no.14, p. 27.

Varley, M. (1991) Stress and reproduction. *Pig News and Information* 12(4), 567–571.

Visscher, P.M., Van der Beek, S. and Haley, C.S. (1996) Marker assisted selection in livestock. *Proceedings of the British Society of Animal Science* (Winter Meeting), Paper no. 52.

Waberski, D., Weitze, K.F., Lietmann, C., Lubbert zur Lage, W., Bortolozzo, F.P., Willmen, T. and Petzoldt, R. (1994a) The initial fertilizing capacity of longterm-stored liquid boar semen following pre- and postovulatory insemination. *Theriogenology* 41, 1367–1377.

Waberski, D., Weitze, K.F., Gleumes, T., Schwarz, M., Willmen, T. and Petzoldt, R. (1994b) Effect of time of insemination relative to ovulation on fertility with liquid and frozen boar semen. *Theriogenology* 42, 831–840.

Waberski, D., Meding, S., Dirksen, G., Weitze, K.F., Leiding, C. and Hahn, R. (1994c) Fertility of long-term-stored boar semen: influence of extender (Androhep and

Kiev), storage time and plasma droplets in the semen. *Animal Reproduction Science* 36, 145–151.

Walker, N. (1980) Pigs per litter depend on the boar. *Agriculture in Northern Ireland* 55, 135–137.

Watson, P.F. (1990) Artificial insemination and the preservation of semen. In: Lamming, G.E. (ed.) *Marshall's Physiology of Reproduction*, 4th edn, Vol. 2, Churchill Livingstone, London, pp. 747–896.

Webb, A.J. (1995) Future challenges in pig genetics. *Animal Breeding Abstracts* 63(10), 731–736.

Wechsler, B., Schmid, H. and Moser, H. (1991) The Stolba family pen for domestic pigs. An ethical management system for breeding and growing pigs. *Tierhaltung* 22, 95pp.

Weiler, U., Claus, R. and Hahn, R. (1988) Light programs for the control of reproduction in AI boars. *Proceedings of the 11th International Congress on Animal Reproduction and AI* (Dublin), 4, Paper no. 419.

Weitze, K.F. (1990) The use of 'long-term extender' in pig AI – a view of the international situation. *Pig News and Information* 11(1), 23–26.

Weitze, K.F. (1991) Long-term storage of extended boar semen. *Proceedings of the 2nd International Conference on Boar Semen Preservation* (Beltsville), pp. 231–253.

Weitze, K.F. and Waberski, D. (1996) Influence of seminal plasma on the timing of ovulation and sperm transport in pigs. *Journal of the British Fertility Society* 1(1), 14A–15A.

Weitze, K.F., Roth, D., Willmen, T., Waberski, D. and Lotz, J. (1990) Advancement of ovulation in the sow related to seminal plasma application before insemination. *Reproduction in Domestic Animals* 25, 61–67.

Weitze, K.F., Stampa, E., Richter, L., Willmen, T. and Waberski, D. (1991) Fertility of frozen boar semen: influence of packaging, number of inseminations and seminal plasma. *Proceedings of the 2nd International Conference on Boar Semen Preservation* (Beltsville), 139–142.

Wekerle, L. (1993) Problems of artificial insemination of pigs in Hungary. *Magyar Allatorvosok Lapja* 48, 162–164.

Westendorf, D.G., Richter, L. and Treu, H. (1975) Zur Tiefgefrierung von Eber-sperma. Laborund Besamungsergebnisse mit dem Hulsenberger Pailletten-Verfahren. *Deutsch Tierartzl Wochenenschrift* 82, 261–267.

Wettemann, R.P., Welty, S. and Bishop, D.K. (1992) An attempt to stimulate sexual behaviour of boars. *Animal Science Research Report*, Oklahoma Agriculture Experimental Station, pp. 410–412.

Whaley, S.L., Britt, J.H., Farin, C.E., Martus, N.S. and Hedgpeth, V.S. (1994) Effect of vitamin A injection on oocyte maturation and early embryonic development in gilts fed high-energy diets before and after mating. *Journal of Animal Science* (Suppl. 1)/*Journal of Dairy Science* (Suppl. 1), p. 78.

White, B.R. and Wheeler, M.B. (1995) Examination of ovulation rate and uterine and fetal interactions in Chinese Meishan, Yorkshire, and reciproacl cross gilts: effects of fetal and maternal genotypes. *Animal Reproduction Science* 39, 147–158.

Wiley, R.E. (1993) Spermicidal effect of latex gloves during semen collection in the boar: a case report. *Swine Health and Production* 1(3), 22–23.

Williams, S., Marotta, E., Lagreca, L. and Vales, L. (1991) Variation in the quality of boar semen due to the frequency of mounting. *Revista de Medicina Veterinaria (Buenos Aires)* 72, 288–294.

Wilson, M.E., Kaminski, M.A., Conley, A.J. and Ford, S.P. (1995) Developmental differences between Meishan and Yorkshire preimplantation conceptuses. *Biology of Reproduction* 52 (Suppl. 1), 158.

Winfield, C.G. (1980) Oestrus detection in pigs. In: Behaviour in Relation to Reproductive Management and Welfare of Farm Animals. *Reviews of Rural Science* 4, 83–86.

Woelders, H. (1991) Overview of in vitro methods for evaluation of semen quality. *Proceedings 2nd International Conference on Boar Semen Preservation* (Beltsville), pp. 145–164.

Wu, M.C. and Dziuk, P.J. (1995) Relationship of length of uterus in prepubertal pigs and number of corpora lutea and fetuses at 30 days of gestation. *Animal Reproduction Science* 38, 327–336.

Xue, J.L., Dial, G.D., Marsh, W.E., Davies, P.R. and Momont, H.W. (1993) Influence of lactation length on sow productivity. *Livestock Production Science* 33, 253–265.

Yen, H.T., Lee, D.N., Yu, I.T. and Chen, S.C. (1991) Factors affecting sow reproduction. 2. Differences between breeds. *Journal of the Chinese Society for Animal Science* 20(1), 19–25.

Youngs, C.R., Christenson, L.K. and Ford, S.P. (1994) Investigations into the control of litter size in swine: III. A reciprocal embryo transfer study of early conceptus development. *Journal of Animal Science* 72, 725–731.

Zak, L.J., Cosgrove, J.R., Aherne, F.X. and Foxcroft, G.R. (1995) Effect of feeding level during lactation upon reproductive function in the primiparous lactating sow. *Journal of Animal Science* 73 (Suppl. 1), 306.

Zink, M.F. and Dielh, J.R. (1984) Efficacy of using vaginal conductivity as an indiactor of optimum time to breed swine. *Journal of Animal Science* 59, 869–874.

2 The Sow's Oestrous Cycle and Associated Events

2.1. Introduction

Of the various farm mammals, the pig is the species which has been influenced to the greatest extent by modern intensive husbandry practices. As noted by Signoret (1980a), permanent housing in controlled environments has resulted in marked changes in a pig's natural social interactions, and this is especially true in reproduction. In contrast to events in the permanently cyclic domestic sow, in the wild pig, sexual activity usually only occurs in the autumn and farrowing at the end of the winter (Kyle, 1995). It has been recorded, however, that such sexual activity can be very well synchronized in wild herds; such simultaneous farrowing would have an obvious survival value in view of the fact that cross-suckling of the young is often observed.

2.1.1. Neuroendocrine mechanisms

When discussing the reproductive behavioural patterns and endocrinology of sexual activity in the female pig, it is as well to remember that, although the domesticated sow faces a situation far removed from that of her wild ancestors, she has to work with the same basic neuroendocrine mechanisms (Fig. 2.1). In this, internal and external stimuli, detected by the central nervous system, are translated by the neuroendocrine system into signals that alter pituitary gland function.

2.2. Oestrus and the Oestrous Cycle

The domestic pig does not usually show evidence of a breeding season, and in the absence of pregnancy the sow can be expected to exhibit oestrus at intervals averaging 21 days (range 19–23 days) throughout the year (McKenzie, 1926; Asdell, 1964). Although the expected average interval between heats is

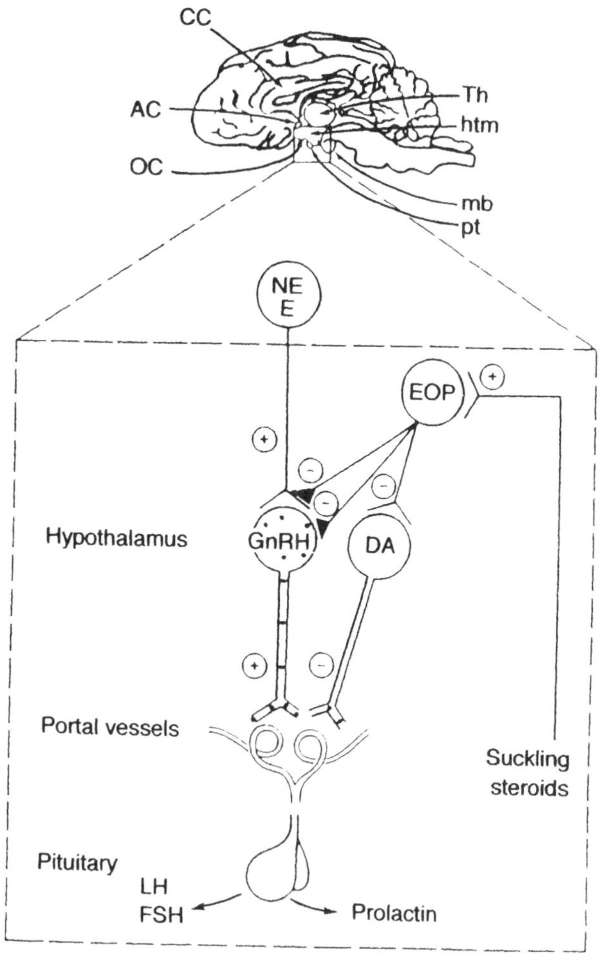

Fig. 2.1. Diagrammatic representation of the pig's neuroendocrine system with suggested nerve network regulating LH, FSH and prolaction secretion. The box represents the hypothalamus (htm)-pituitary gland (pt) unit. AC = anterior commissure, CC = corpus callosum, DA = dopamine, E = epinephrine (adrenaline), EOP = endogenous opioid peptide, GnRH = gonadotrophin releasing hormone, mb = mammillary body, NE = norepinephrine (noradrenaline), OC = optic chiasma, Th = thalamus (from Kraeling and Barb, 1990).

normally 21 days, Dziuk (1977) showed that when a large number of oestrous cycles is recorded in mated gilts, there are two peaks in the frequency of these intervals – one occurs near 21 days and the other at 26 days. It is noteworthy that similar 26-day intervals have been reported in mated gilts possessing only one to four embryos (Polge *et al.*, 1966); the indications are that a small number of embryos prolongs the normal 21-day interval but does not initiate a full-term pregnancy.

2.2.1. Variability in cycle length

The occurrence of two phases of return to oestrus in sows on commercial units in the UK was also recorded in a study of 2472 female pigs (Glossop and Foulkes, 1988); one phase peaked at 20.7 days and the second at 26.5 days. It was suggested that two mechanisms might be operating to account for this; the first peak derived from a failure of fertilization and the second from embryonic loss at a later stage. The stockperson should be aware of such occurrences.

As already noted, oestrous cycles are generally thought of as occurring throughout the year without any obvious seasonality (Asdell, 1964; Dziuk, 1991). However, there is evidence of a tendency, at least in some sows, to show cyclical breeding activity rather less consistently during the late summer (Stork, 1979; Hurtgen and Leman, 1980). Studies at Oxford produced some evidence that sows culled from outdoor pig herds for failure to breed in summer may be intrinsic seasonal breeders (Bassett *et al.*, 1994); they found evidence of declining LH secretion in such sows during late spring/summer which may contribute towards the genesis of summer infertility.

Although the duration of the normal porcine oestrous cycle remains relatively constant at 21 days, there are likely to be minor individual and breed variations; it is generally held that cycle length should not be affected by factors such as feeding and management. According to Asdell (1964), cycles tend to be a little shorter in gilts than in mature sows.

2.2.2. Oestrus in the pig

The start of the heat period is characterized by gradual changes in the pig's behaviour which may not always be easily recognized (increased restlessness, reduced appetite, mounting other animals, male-like sexual behaviour, lordosis response) and physiological changes that may also go undetected (swelling and colour changes of the vulva, occasionally a cloudy mucus discharge). The duration of sexual receptivity in the normal sow has been reported by many workers. McKenzie and Miller (1930) gave the average as 40–46 h; more than 60 years later, Anderson (1993) recorded a figure of 40–60 h. Many authors simply report that oestrus lasts 2–3 days, but that variation from 1 to 4 days is commonplace.

2.2.3. Precopulatory behaviour in pigs

The development of sexual behaviour in pigs has been reviewed by Ford (1990); according to whom, the late development and definite role of the olfactory system in sexual behaviour in the pig identify this species as an excellent model for studies of mechanisms associated with the development and regulation of mammalian sexual behaviour. On the more practical level, various observers have recorded courtship behaviour in pigs (Fig. 2.2). In Japan, for example, Tanida *et al.* (1989) distinguished seven stages of courtship behaviour in boars: sniffing, head to head, nosing, following the sow, chin-resting, mounting and copulation. The authors used sequence analysis to determine the pattern of the boar's behaviour before mating. The most frequently observed transition was from head-to-head contact with the sow to nosing; nosing was at the centre of the courtship behaviour pattern.

The mounting response of the boar appears to be produced mainly by the immobilization of the sow. Signoret (1980) observed that the overall shape of the body of the sow and its immobility are the major stimuli; other characteristics displayed by the female are likely to be of minor importance. This would account for the rapid response observed in both sexually experienced and inexperienced boars to the simplified dummies that are used for semen collection.

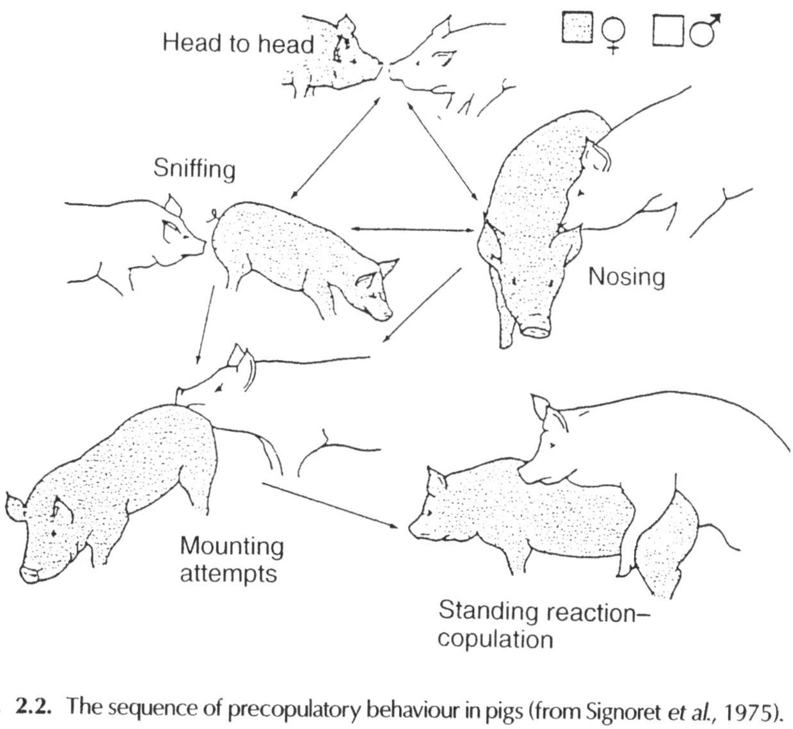

Fig. 2.2. The sequence of precopulatory behaviour in pigs (from Signoret *et al.*, 1975).

2.2.4. Using the boar in oestrus detection

The only certain method of detecting oestrus in the pig is to ensure that the female is in daily contact with a boar of good libido in an environment as free from distractions as possible. It is preferable to take sows to the boar rather than the other way round; away from his home pen, investigatory and other forms of behaviour may take precedence over sexual behaviour. The boar pen must be located near the sows for this procedure to be easily carried out; proximity of the boar to sows, in any event, may have a positive effect on the full display of oestrous behaviour. It should be noted that there have been reports showing that group size and stocking density may influence the efficiency of oestrus detection in the presence of the boar (Hemsworth and Barnett, 1990); the need to examine and clarify the optimal social and spatial conditions for group-housed female pigs has been emphasized.

2.2.5. Oestrus detection in the absence of the boar

In the absence of the boar, the skill and experience of the stockperson is vital if visual and behavioural signs are to be interpreted accurately. Changes in the appearance of the external genitalia (vulva becomes oedematous and its colour changes from pink to bright red) may be observed 2–6 days before the onset of oestrus; such changes have usually subsided before the period of sexual receptivity. The changes in gilts are generally much more marked than in older sows (Winfield, 1980). Oestrous sows may be observed mounting or allowing themselves to be mounted; however, mounting behaviour is not shown by sufficient sows to constitute an effective means of detection. There is also a danger that similar behaviour may be shown when sows are forming new social orders in groups after weaning (dominance–subordinance traits). A report from Denmark by Pedersen *et al.* (1993), dealing with postweaned grouped sows, recorded evidence of the effects of aggression on oestrous behaviour; sows receiving the highest amounts of aggression showed significantly less social behaviour during oestrus than other sows. The difference in oestrous behaviour between dominant and subordinant sows suggested that the detection of oestrus may be more difficult in subordinate animals.

A behavioural characteristic frequently used by the stockperson to detect oestrus in the pig is the 'riding test'; however, many gilts and sows may not exhibit this response in the absence of the boar. In such females, the response may be shown only in the presence of stimuli associated with the boar; among these stimuli is the mating song and the sexual odour of the boar (Table 2.1). The main sources of these odours, identified as androstenone and hydroxyandrostene, are the preputial and salivary glands, respectively. The preputial secretions are particularly effective in inducing the immobilization reflex; boars from which the submaxillary salivary glands had been removed were unable to elicit a full immobilization reflex in oestrous pigs (Perry *et al.*, 1972).

It is noteworthy that olfactory sensitivity in pigs may be sexually

Table 2.1. Effect of various stimuli facilitating the immobilization reflex in the oestrous sow (from Signoret, 1971).

Conditions of stimulation	% of oestrous sows standing
With the boar	100
In the presence of the boar, response to dorsal stimulation by the experimenter	97
the boar invisible (hessian screen)	90
In the absence of the boar dorsal stimulation by the experimenter only	48
plus broadcasting of boar's 'courting' grunts	75
plus boar's preputial secretion	80
plus spray of 5-α-androstene	81

dimorphic. Dorries *et al.* (1995) examined the olfactory sensitivity of adult boars and sows and in castrate pigs. They found evidence of a sex difference in sensitivity to the odour of androstenone; female pigs were much more sensitive than intact males.

2.2.6. Combined detection techniques

The individual testing of sows in a boar pen, especially gilts, may be regarded as impracticable and the detection of oestrus in the absence of the boar to be less than reliable. A combined technique (Winfield, 1980) may be more useful. In this, a preliminary examination for oestrus is undertaken by the stockperson followed by the introduction of selected sows to the boar pen. Whatever system of oestrus detection is chosen, the motivation and skill of the stockperson is likely to have a considerable influence on its effectiveness.

2.2.7. Vaginal cell cytology

There may be circumstances in which it is necessary to resort to unusual methods of detect oestrus in pigs. A report by Rodgers *et al.* (1993) deals with oestrus detection in Yucatan miniature pigs by way of vaginal smears stained with haematological stain; epithelial cells were categorized as superficial, large intermediate, small intermediate or parabasal. Epithelial cells were at their peak during oestrus, decreasing markedly during dioestrus and increasing again during pro-oestrus; the combined superficial plus large intermediate cell counts were significantly higher during oestrus than at other stages of the cycle. The authors concluded that examination of daily vaginal smears could be used in oestrus detection.

2.2.8. Duration of oestrus

Gilts frequently fail to show oestrus for more than one day, whereas sows usually show sexual receptivity for two days or longer. The pubertal oestrus is usually shorter than later ones; Anderson (1993) quotes values of 47 and 56 h, respectively. Postpartum oestrus has been recorded in sows in the early days (2–5) after farrowing (Warnick *et al.*, 1950; Burger, 1952; Baker *et al.*, 1953); in view of the fact that there is no marked development in the ovaries at this time, it is presumed that an extra-ovarian source of oestrogen is responsible for this anovulatory heat. It would appear that this oestrus may be the result of the action of circulating oestrogens derived from fetuses rather than from the ovaries (Holness and Hunter, 1975).

2.2.9. Ovulation

Ovulation is spontaneous in the pig and usually occurs about 40 h after the onset of oestrus when the heat period lasts about 2 days; when the duration is longer than 2 days, it occurs when about 75% of the heat period has elapsed. In normal cyclic gilts, the interval from release of preovulatory LH to ovulation is about 40 h, the release of the LH being at the onset of oestrus (Niswender *et al.*, 1970). The LH surge induces ovulation, the differentiation of follicular cells (primarily granulosa but with some theca) and the formation of corpora lutea (CL)). The number of oocytes released from the ovaries varies, usually being within the range 10–24, depending on factors such as age, breed, parity and nutrition; the corpora lutea form as solid masses of luteal cells within 6–8 days. At the pubertal oestrus, ovulation rate tends to be lower than at later heats; some authors have recorded a marked increase of up to three ovulations at the second heat and a further, less substantial increase, may be recorded at the third period (Warnick *et al.*, 1951).

Duration of ovulation

The duration of ovulation, i.e. time taken to shed all the oocytes, has been reported to vary from about 1 to 6 h (Betteridge and Raeside, 1962; Du Mesnil du Buisson and Signoret, 1970; Soede and Kemp, 1993) but frequent stimulation by the boar during pro-oestrus and the early hours of the heat period may concentrate the ovulation process (Signoret *et al.*, 1972; Hunter, 1977). It is also clear that ovulations may occur several hours earlier in mated than in unmated animals (Signoret *et al.*, 1972); this is believed to be due to the presence of oestrogen and an unidentified peptide fraction in seminal plasma (Weitze and Waberski, 1996).

In the Netherlands, Soede and Kemp (1993), examining factors that may be involved in early embryonic diversity in pigs, used real-time ultrasound scanning to monitor the ovulation process. They concluded that a difference of up to 3 h in the duration of ovulation was unlikely to be an important determinant of variation among sows in embryonic diversity.

2.3. Hormones of the Oestrous Cycle

The published literature on the levels of steroid hormones that operate during the oestrous cycle of the pig suggests many similarities with cattle and sheep (Fig. 2.3). Oestrogen level rises during the follicular phase of the cycle, triggering the release of the preovulatory surge of luteinizing hormone (LH) at the onset of oestrus. There is just one sharp LH peak at oestrus; the level of the gonadotrophin remains low during the remainder of the cycle.

2.3.1. Progesterone in the oestrous cycle

After ovulation, there is a rapid proliferation of cells (primarily granulosa but with a few theca cells) lining the collapsed follicle wall; these cells, as a result of the preovulatory LH surge, become luteinized to form luteal tissue, a process which involves rapid functional changeover from oestrogen to progesterone production and morphological differentiation into more typical steroidogenic cells. The process of luteinization proceeds rapidly, so that by 6–8 days the corpus luteum has become a ball of luteal cells with an overall diameter of 8–11 mm. By that time, healing of the follicle rupture point by stromal connective tissue has usually occurred, but reorganization of the blood-coagulum-filled central cavity takes longer; eventually, and certainly if

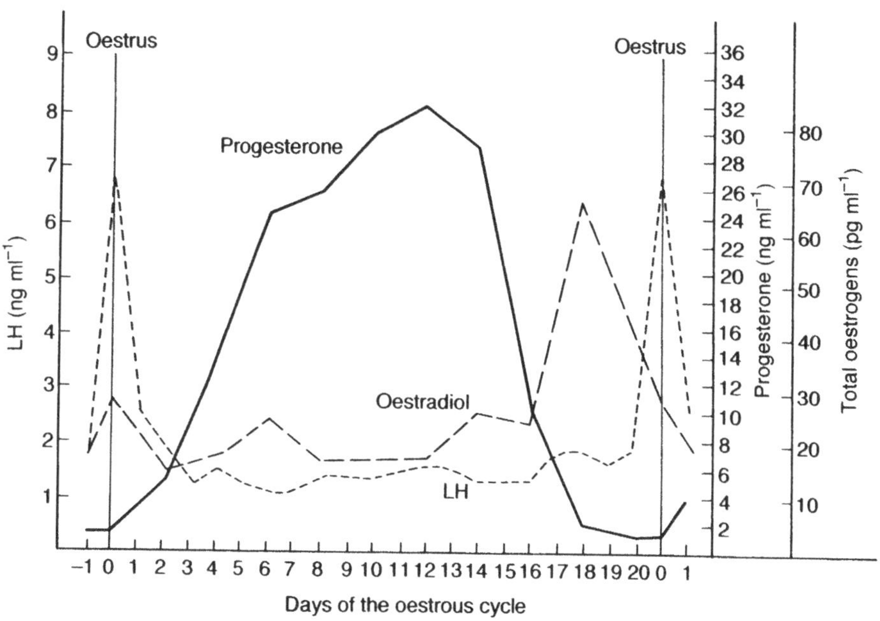

Fig. 2.3. Diagrammatic representation of hormone levels during the oestrous cycle of the pig (from Hansel *et al.*, 1973).

the pig becomes pregnant, the central cavity is obliterated by connective tissue and luteal cell hypertrophy. In secretory terms, progesterone levels are low at oestrus but begin to increase by day 2 and reach peak values by days 8 to 12; these levels are a reflection of the morphological and ultrastructural changes occurring in the luteal cells. Soon after the mid-cycle stage, there is a precipitous decline in the level of progesterone.

Relaxin in the porcine corpus luteum

The occurrence and localization of relaxin in luteal tissue during the oestrous cycle of the pig has been studied by various workers. In Hawai, Bagnell *et al.* (1989) showed that relaxin is present in the porcine corpus luteum during its formation and functional lifespan; it appeared that the presence of relaxin in luteinizing granulosa cells after ovulation was associated with cell luteinization.

2.3.2. Maintenance of the corpora lutea

Once initiated by the preovulatory surge of LH, subsequent formation and function of the corpus luteum in the sow is believed to proceed independently of further gonadotrophic hormone support, for at least the duration of its oestrous cycle activity. If sows are hypophysectomized on the first day of oestrus, corpora lutea are formed and remain active for about 15 days (Anderson *et al.*, 1967). Provided that ovulation takes place, failure of a ruptured follicle to form a corpus luteum would seem to be a remote possibility. Premature failure of corpora lutea during the oestrous cycle is also unlikely because they are apparently insensitive to luteolytic factors such as $PGF_{2\alpha}$ until they are least 10 days old (Gleeson, 1974), which only leaves about 4 days before the onset of normal cyclic regression.

Plasma progesterone levels have been recorded as reaching a peak in the region of 25–35 ng ml^{-1} at about days 8–10 of the cycle; the level is maintained until day 14 or 15 when a steady decline in progesterone level begins, with the concentration of the steroid declining to preovulatory levels by the last days of the cycle.

Luteotrophic action of oestrogens

During the luteal phase of the cycle, oestrogens are very luteotrophic (Dziuk, 1991); an injection or oral administration of an oestrogen can prolong the lifespan of porcine corpora lutea for several weeks. The ingestion of mycotoxins such as zearalenone (from mouldy feed) will also extend the lifespan of corpora lutea. According to Dial and Britt (1986) cyclic gilts exposed to zearalenone often enter into a pseudopregnant condition, with the prolonged maintenance of the corpora lutea and continued elevations in circulating levels of progesterone. Such conditions may sometimes lead the stockperson to conclude that gilts are pregnant when they are pseudopregnant.

2.3.3. Prostaglandins and luteolysis

About 14–15 days after its formation, the lifespan of the porcine corpora lutea is decided; the lifespan will be extended in the pregnant animal whereas in the non-pregnant cyclic pig there is rapid regression of the corpora lutea and development of a new wave of preovulatory follicles. Morphological regression of the corpora lutea is mirrored by cessation of luteal function, the levels of plasma progesterone in the circulation falling rapidly within a period of 1 or 2 days. The mechanisms involved in terminating or (in the case of the pregnant sow) extending the lifespan of the corpora lutea are not fully understood. The regression of the corpora lutea is associated with the presence of the uterus (Puglisi *et al.*, 1978); if the organ is removed this will permit the corpora lutea to persist. There is an increase in the level of $PGF_{2\alpha}$ in the utero-ovarian vein between days 12 and 16 of the oestrous cycle which leads to luteolysis (Bazer *et al.*, 1982). The reason for the early initiation of prostaglandin release in the sow is not apparent, because at the time it is first occurring it is difficult to induce regression in corpora lutea with exogenous prostaglandin. There are various interactions between oestrogens, prostaglandin and other factors in the luteolytic process which have only become apparent in recent years.

During the process of luteolysis in pigs, corpora lutea are invaded by macrophages which produce tumour necrosis factor (TNF). Results reported by Wuttke *et al.* (1995) in Germany, have indicated that TNF inhibits oestradiol production and due to this inhibition an indirect oestradiol-mediated luteotrophic effect of $PGF_{2\alpha}$ is prevented and its direct luteolytic (i.e. its progesterone inhibitory effect) becomes dominant. The authors conclude that both TNF and $PGF_{2\alpha}$ act synergistically during the process of luteolysis in the pig.

2.4. Action of Gonadotrophins during the Cycle

2.4.1. Release of LH

Niswender *et al.* (1970) were the first to report on a radioimmunoassay (RIA) procedure for LH in the pig and to publish data on plasma levels. Brinkley (1981) noted that the most dramatic feature of the sow's gonadotrophin profile is the preovulatory LH surge, which begins to rise in the early hours of oestrus, reaching a peak of 8ng ml^{-1} or more, before returning to basal levels of 1–2ng ml^{-1} some 43h later; the long duration of the LH surge in the sow is in contrast to the situation in the cow and ewe.

The LH surge at the onset of oestrus initiates both maturation of the oocyte in the preovulatory follicle and rupture of the follicle wall. In the follicle destined for ovulation, the primary oocyte commences its first meiotic division about 20h after the LH peak, and proceeds with its reduction division, with formation of the first polar body about 2h before ovulation (Hunter and Polge, 1966; Hunter, 1977). The duration of meiotic maturation is substantially longer in pigs (42–44h) than in cattle, sheep and goats (24h).

The preovulatory surge of LH in the pig is believed to be triggered off by the increasing concentration of oestradiol secreted by the follicles that are destined to ovulate during the sow's heat period. The interval between peak oestradiol concentration and peak LH concentration was recorded by Soede *et al.* (1994) as 11 ± 4h. Evidence supports the existence of an LH surge generator in the hypothalamus of the pig. Unlike in cows and ewes in which oestradiol stimulates the LH surge by acting both on the central nervous system and the anterior pituitary, oestradiol apparently stimulates the preovulatory surge in pigs by acting primarily on the central nervous system (Kraeling and Barb, 1990).

2.4.2. Pulsatile release of LH

The presence of LH pulses at certain stages of the pig's oestrous cycle was discussed by Brinkley (1981), who found these to be of 60–80 minutes duration during the mid-luteal phase; in such pulses, the LH concentrations could rise to levels almost as high as those found in the preovulatory surge. In the follicular phase of the cycle, on the other hand, there appeared to be an absence of pulses and this persisted until the onset of the preovulatory surge. During the long period of the preovulatory surge of LH, apparently pulses were not observed. As in cattle and sheep, pulsatile LH secretion in pigs probably stems from pulsatile secretion of gonadotrophin releasing hormone (GnRH) from the hypothalamus; oestrus and ovulation can be induced by pulsatile intravenous injections of GnRH (Lutz *et al.*, 1985).

Pulsatile release of LH can be abolished by hypophysial stalk transection (Kesner *et al.*, 1989) as well as by passive immunization against GnRH (Esbenshade *et al.*, 1986), supporting the view that there is a 'pulse generator' within the central nervous system. As shown by Estienne *et al.* (1989), progesterone will not suppress LH secretion in response to pulsatile administration of GnRH in the hypophysial-stalk sectioned pig, suggesting that the negative feedback action of this steroid on pulsatile LH secretion in pigs occurs at the level of the hypothalamus rather than at the pituitary.

Oestradiol control of LH

It is known that exogenous oestradiol can be used to inhibit LH secretion in the ovariectomized sow and in prepubertal gilts. There is also the fact that tonic LH concentrations are lowest during the follicular phase of the cycle and no LH pulses are exhibited at this time of high oestradiol concentrations; such evidence indicated that follicular oestradiol suppresses the appearance of LH pulses in the late stages of the cycle.

2.4.3. Follicular development during the porcine cycle

The use of scanning by real-time ultrasonics has enabled the pattern of follicular development during the oestrous cycle to be described for the cow (Fortune *et al.*, 1988; Savio *et al.*, 1988) and sheep (Driancourt, 1991). These studies identified the emergence of two or three follicular waves during the cycle and the continuation of follicular wave development during early pregnany. A review by Fortune (1994) suggested that there are probably two different patterns of development of large follicles in mammals. In one pattern, as shown in humans, rats and pigs, the development of ovulatory-sized follicles is suppressed, except during the follicular phase of the cycle; selection of preovulatory follicles is believed to occur between days 14 and 16 of the cycle (Clark *et al.*, 1982; Foxcroft and Hunter, 1985; Foxcroft, 1987). In the other pattern, as shown in cattle, sheep and horses, development of follicles to ovulatory or near-ovulatory size is not confined to the follicular phase, but occurs throughout the cycle. In prepubertal gilts, there have been some indications that the morphology of the ovaries undergoes continuous changes, suggestive of follicular waves (Bolamba *et al.*, 1994).

In Ireland, Ryan *et al.* (1994) found no evidence of follicular waves in either pregnant or non-pregnant sows; the total follicle population in the ovaries of cyclic animals remained the same from day 4 until day 20. In the cyclic pig, they observed no evidence of a change in size of follicles until day 15; from day 15 until day 20 of the cycle, potential ovulatory follicles were recruited from a pool of small follicles. This recruitment of follicles is believed to be related to the elevation of utero-ovarian plasma vein concentrations of $PGF_{2\alpha}$ which is known to occur between days 12 and 16 of the cycle, leading to luteolysis (Bazer *et al.*, 1982). The regression of the corpora lutea and the consequent decrease in progesterone levels presumably allows stimulation of the pool of small follicles by gonadotrophins (Ainsworth *et al.*, 1980; Huff and Esbenshade, 1992).

During the luteal phase of the porcine oestrous cycle, 40–50 follicles measuring 3–6 mm in diameter are likely to be present on the surface of the ovaries; of these 50% are known to be atretic. Healthy non-atretic follicles of 3 mm and above grow at the rate of 1 mm day^{-1}. As observed by Guthrie *et al.* (1995), in a study of follicular atresia, the follicle population is maintained by a physiological equilibrium between follicle proliferation and apoptotic cell death. This equilibrium is disrupted during the follicular phase of the cycle, when a few follicles undergo preovulatory maturation and ovulation; the remainder die and are not replaced after ovulation. The authors found that the percentage of medium-sized follicles undergoing atresia (defined by granulosa cell apoptosis) increased from 6% to 50% between days 5 and 7.

The first period of follicular development after ovulation in the pig results in a tenfold increase in the number of medium-sized (3–6 mm diameter) follicles between days 3 and 8 of the oestrous cycle. A study by Guthrie and Cooper (1995) looked for evidence of waves of follicular growth and atresia between days 7 and 15 of the cycle; their results supported the view that there is continuous growth and atresia of ovarian follicles during the luteal phase with no evidence of follicle dominance or discernible waves of growth and atresia.

In Hungary, Ratky *et al.* (1995) used endoscopy to observe ovarian follicle development during the first few oestrous cycles in gilts; they recorded that follicle recruitment had started by day 13. Follicles of 4 mm diameter on day 14 matured to the preovulatory stage whereas follicles of 2–3 mm became atretic; preovulatory follicles grew at approximately 1 mm per day between days 14 and 19. The same workers recorded a significant increase in the average ovulation rate during the first three heat periods; the number of ovulations at the first, second and third heats averaged 8.0, 9.5 and 11.0, respectively.

2.4.4. Gonadotrophins and follicle development

The vesicular follicles in the ovaries of the pig do require gonadotrophins for their support; the presence of midluteal plasma LH pulses would seem to be associated with the process of atresia and follicular maturation which is found to be operating during the later stages of the cycle. In the follicular phase of the pig's cycle, those vesicular follicles which are destined to ovulate grow from about 4–5 mm diameter on day 15 to an ovulatory diameter of 9–11 mm.

As with LH, plasma follicle stimulating hormone (FSH) concentrations are known to be low during the follicular phase and it is also evident that oestrogens will suppress FSH levels. The period of low plasma FSH occurs during the highest and most prolonged period of oestradiol concentration in the cycle and is followed by the LH surge, the main FSH surge and the onset of oestrus. It is in the follicular phase that follicles mature, granulosa cells in these follicles proliferate rapidly and produce the oestradiol, plasma levels of this steroid increasing to a peak of 60–90 pg ml^{-1} on about day 19 of the cycle (Hunter *et al.*, 1972). It would seem that oestradiol initially suppresses both plasma LH and FSH levels, but later induces the preovulatory surge of gonadotrophin.

In the immediate postovulatory period, the abrupt decline in circulating oestradiol concentrations is believed to permit a marked increase in the secretion of FSH that continues for 2–3 days after oestrus (Dial and Britt, 1986). It has been suggested that this increased release of FSH may be in response to the disappearance of follicular sources of either oestradiol or inhibin, both of which are known to exert a negative feedback effect on the release of this gonadotrophin.

2.4.5. Prolactin and follicle development

The role of prolactin in the oestrous cycle of the pig remains unclear, but secretion of the hormone is known to be controlled by hypothalamic factors

(Kraeling and Barb, 1990), the most notable of which is the prolactin-inhibitory factor (dopamine) and the prolactin-releasing activity of thyroid-stimulating hormone-releasing hormone (TRH). In terms of ovarian action, the effect of prolactin on the developing follicle is closely related to the size of follicle. In small follicles (1–2 mm), progesterone synthesis is inhibited by prolactin; in larger follicles (>6 mm) progesterone production is stimulated (Ellendorff and Parvizi, 1987). It is possible that oestrogens, which are present in large quantities in the larger follicles are responsible for this biphasic action.

2.4.6. Paracrine regulators of follicle development in pigs

Follicular growth is not only a matter of changing concentrations of FSH and LH; their control of the follicular microenvironment is believed to be permissive. According to Tonetta and diZerega (1990) the developing responsiveness of follicles to stimulation by FSH and LH results from changes in the production of and alterations in follicular sensitivity to intraovarian paracrine and/or autocrine factors (inhibin, activin, follistatin, follicle regulatory protein, glycosaminoglycans and growth factors).

2.5. References

Ainsworth, L., Tsang, B.K., Downey, B.R., Marcus, G.J. and Armstrong, D.T. (1980) Inter-relationships between follicular fluid steroid levels, gonadotropic stimuli, and oocyte maturation during preovulatory development of porcine follicles. *Biology of Reproduction* 23, 621–627.

Anderson, L.L. (1993) Pigs. In: Hafez, E.S.E. (ed.) *Reproduction in Farm Animals*, 6th edn. Lea & Febiger, Philadelphia, pp. 343–360.

Anderson, L.L., Dyck, G.W., Mori, H., Henricks, D.M. and Melampy, R.M. (1967) Ovarian function in pigs following hypophysial stalk transection or hypophysectomy. *American Journal of Physiology* 212, 1188.

Asdell, S.A. (1964) *Patterns of Mammalian Reproduction*, 2nd edn. Cornell University Press, Ithaca, USA, 670 pp.

Bagnell, C.A., Ayau, E., Downey, B.R., Tsang, B.K. and Ainsworth, L. (1989) Localization of relaxin during formation of the porcine corpus luteum. *Biology of Reproduction* 40, 835–841.

Baker, L.M., Woehling, H.L., Casida, L.E. and Grummer, R.H. (1953) Occurrence of oestrus in sows following parturition. *Journal of Animal Science* 12, 33–38.

Bassett, J.M., Bray, C.J. and Sharpe, C.E. (1994) Melatonin administration and seasonal changes in plasma luteinizing hormone (LH) and prolactin (PRL) in a selected population of domestic sows kept outdoors without boars. *Journal of Reproduction and Fertility*, Abstract Series no. 14, pp. 7–8.

Bazer, F.W., Geisert, R.D., Thatcher, W.W. and Roberts, R.M. (1982) Endocrine vs. exocrine secretion of PG F2-alpha in the control of pregnancy in swine. In: Edqvist, L.E. and Kindahl, H. (eds) *Prostaglandins in Animal Reproduction* II. Elsevier Science, Amsterdam, pp. 115–132.

Betteridge, K.J. and Raeside, J.I. (1962) Observation of the ovary by peritoneal

cannulation in pigs. *Research in Veterinary Science* 3, 390–398.
Bolamba, D., Matton, P., Estrada, R. and Dufour, J.J. (1994) Ovarian follicular dynamics and relationship between ovarian types and serum concentrations of sex steroids and gonadotrophin in prepubertal gilts. *Animal Reproduction Science* 36, 291–304.
Brinkley, H.J. (1981) Endocrine signalling and female reproduction. *Biology of Reproduction* 24, 22–43.
Burger, J.F. (1952) Sex physiology of pigs. *Ondestepoort Journal of Veterinary Research*, Suppl. 2, 1–218.
Clark, J.R., Brazier, S.G., Wiginton, L.M., Stevenson, G.R. and Tribble, L.F.(1982) Time of ovarian follicle selection during the porcine estrous cycle. *Theriogenology* 18, 697–709.
Dial, G.D. and Britt, J.H. (1986) The clinical endocrinology of reproduction in the pig. In: Morrow, D.A. (ed.) *Current Therapy in Theriogenology.* W.B. Saunders, Philadelphia, pp. 905-911.
Dorries, K.M., Adkins-Regan, E. and Halpern, B.P. (1995) Olfactory sensitivity to the pheromone, androstenone, is sexually dimorphic in the pig. *Physiology and Behaviour* 57, 255–259.
Driancourt, M.A. (1991) Follicular dynamics in sheep and cattle. *Theriogenology* 35, 207–220.
Du Mesnil du Buisson, F. and Signoret, J.P. (1970) Reproductive physiology and AI in pigs. *Veterinary Record* 87, 562–568.
Dziuk, P.J. (1977) Reproduction in pigs. In: Cole, H.H. and Cupps, P.T. (eds) *Reproduction in Domestic Animals,* 3rd edn. Academic Press, New York, pp. 456–476.
Dziuk, P.J. (1991) Reproduction in the pig. In: Cupps, P.T. (ed.) *Reproduction in Domestic Animals,* 4th edn. Academic Press, New York, pp. 471–489.
Ellendorff, F. and Parvizi, N. (1987) Neuroendocrine control of follicular development and function in the sow. In: Roche, J.F. and O'Callaghan, D. (eds) *Follicular Growth and Ovulation Rate in Farm Animals.* Martinus Nijhoff, Dordrecht, pp. 179–189.
Esbenshade, K.L., Vogel, M.J. and Traywick, G.B. (1986) Clearance rate of luteinizing hormone and follicle stimulating hormone from peripheral circulation in the pig. *Journal of Animal Science* 62, 1649–1653.
Estienne, M.J., Kesner, J.S., Kraeling, R.R., Rampacek, G.B. and Barb, C.R. (1989) Luteinizing hormone secretion in hypophysial stalk-transected pigs given progesterone and pulsatile gonadotropin-releasing hormone. *Proceedings of the Society of Experimental Biology and Medicine* 190, 14–17.
Ford, J.J. (1990) Differentiation of sexual behaviour in pigs. *Journal of Reproduction and Fertility Suppl.* 40, 311–321.
Fortune, J.E. (1994) Ovarian follicular growth and development in mammals. *Biology of Reproduction* 50, 225–232.
Fortune, J.E., Sirios, J. and Quirk, S.M. (1988) The growth and differentiation of ovarian follicles during the bovine oestrous cycle. *Theriogenology* 29, 95–109.
Foxcroft, G.R. (1987) Follicular dynamics and its endocrine control in the sow. In: Roche, J.F. and O'Callaghan, D. (eds) *Follicular Growth and Ovulation Rate in Farm Animals.* Martinus Nijhoff, Dordrecht, pp. 191–205.
Foxcroft, G.R. and Hunter, M.G. (1985) Basic physiology of follicular maturation in the pig. *Journal of Reproduction and Fertility, Suppl.* 33, 1–19.
Gleeson, A.R. (1974) Luteal function in the cyclic sow after infusion of prostaglandin F2-alpha through a uterine vein. *Journal of Reproduction and Fertility* 36, 487–488.

Glossop, C.E. and Foulkes, J.A. (1988) Occurrence of two phases of return to oestrus in sows on commercial units. *Veterinary Record* 122, 163–164.

Guthrie, H.D. and Cooper, B.S. (1995) Granulosa cell apoptosis, follicular fluid hormones, and circulating gonadotropins during the mid-luteal phase of the estrous cycle in pigs. *Biology of Reproduction* 52 (Suppl. 1), 150.

Guthrie, H.D., Grimes, R.W., Cooper, B.S. and Hammond, J.M. (1995) Follicular atresia in pigs: measurement and physiology. *Journal of Animal Science* 73, 2834–2844.

Hansel, W., Concannon, P.W. and Lukaszewska, P.H. (1973) Oestrous cycle of the pig. *Biology of Reproduction* 8, 222.

Hemsworth, P.H. and Barnett, J.L. (1990) Behavioural responses affecting gilt and sow reproduction. *Journal of Reproduction and Fertility, Suppl.* 40, 343–354.

Holness, D.H. and Hunter, R.H.F. (1975) Post-partum oestrus in the sow in relation to the concentration of plasma oestrogens. *Journal of Reproduction and Fertility* 45, 15–20.

Huff, B.G. and Esbenshade, K.L. (1992) Induction of follicular development and estrus in prepubertal gilts with gonadotropins. *Animal Reproduction Science* 27, 183–194.

Hunter, R.H.F. (1977) Physiological factors influencing ovulation, fertilization, early embryonic development and establishment of pregnancy in pigs. *British Veterinary Journal* 133, 461–470.

Hunter, R.H.F. and Polge, C. (1966) Maturation of follicular oocytes in the pig after injection of HCG. *Journal of Reproduction and Fertility* 12, 525–531.

Hunter, R.H.F., Hall, J.P., Cook, B. and Taylor, P.D. (1972) Oestrogens and progesterone in porcine peripheral plasma before and after induced ovulation. *Journal of Reproduction and Fertility* 31, 499–501.

Hurtgen, P.J. and Leman, A.D. (1980) Seasonal influence on the fertility of sows and gilts. *Journal of American Veterinary Medical Association* 177, 631–635.

Kesner, J.S., Estienne, M.J., Kraeling, R.R. and Rampacek, G.B. (1989) Luteinizing hormone and prolactin secretion given estradiol and pulsatile gonadotropin releasing hormone in hypophysial stalk-transected pigs. *Neuroendocrinology* 49, 502–508.

Kraeling, R.R. and Barb, C.R. (1990) Hypothalamic control of gonadotrophin and prolactin secretion in pigs. *Journal of Reproduction and Fertility* Suppl. 40, 3–17.

Kyle, R. (1995) Wild boar in Britain: a new farming enterprise. *State Veterinary Journal* 5(2), 10–12.

Lutz, J.B., Rampacek, G.B. and Kraeling, R.R. (1985) Induction of ovulation in the prepubertal gilt by pulsatile administration of gonadotropin releasing hormone. *Domestic Animal Endocrinology* 2, 61–65.

McKenzie, F.F. (1926) The normal estrous cycle in the sow. *Missouri Agricultural Experiment Station Research Bulletin* no. 86.

McKenzie, F.F. and Miller, J.C. (1930) Length of the period of heat in gilts and sows and the length of the interval between heat periods. *Missouri Agricultural Experiment Station Research Bulletin* no. 285.

Niswender, G.D., Reichert, L.E. and Zimmerman, D.R. (1970) RIA of serum levels of LH throughout the estrous cycle in pigs. *Endocrinology* 37, 576–580.

Pedersen, L.J., Rojkittikhun, T., Einarsson, S. and Edqvist, L.E. (1993) Postweaning grouped sows: effects of aggression on hormonal patterns and oestrus behaviour. *Applied Animal Behaviour Science* 38, 25–39.

Perry, G.C., Patterson, R.L.S. and Stinson, G.C. (1972) Submaxillary salivary gland involvement in porcine mating behaviour. *Proceedings of the 7th International Congress Animal Reproduction and AI (Munich)*, pp. 396–399.

Polge, C., Rowson, L.E.A. and Chang, M.C. (1966) The effect of reducing the number of embryos during early stages of gestation in the maintenance of pregnancy in the pig. *Journal of Reproduction and Fertility* 12, 395–397.

Puglisi, T.A., Rampecek, G.B. and Kraeling, R.R. (1978) Corpus luteum function following subtotal hysterectomy in the prepubertal gilt. *Journal of Animal Science* 46, 707–710.

Ratky, J., Brussow, K.P. and Hunter, M.G. (1995) Endoscopic studies of ovarian follicle development during the oestrus cycle in Hungarian Large White gilts. *Archiv fur Tierzucht* 38, 427–435.

Ryan, D.P., Yaakub, H., Harrington, D. and Lynch, P.B. (1994) Follicular development during early pregnancy and the estrous cycle of the sow. *Theriogenology* 42, 623–632.

Savio, J.D., Keenan, L., Boland, M.P. and Roche, J.F. (1988) Pattern of growth of dominant follicles during the oestrous cycle in heifers. *Journal of Reproduction and Fertility* 83, 663–671.

Signoret, J.P. (1971) The mating behaviour of the sow. In: Cole, D.J.A. (ed.) *Pig Production*. Butterworths, London, pp. 295–313.

Signoret, J.P. (1980a) Endocrine basis of reproductive behaviour in female domestic mammals. *Reviews of Rural Science* 4, 3–9.

Signoret, J.P. (1980b) Mating behaviour of the pig. *Reviews of Rural Science* 4, 75–78.

Signoret, J.P., du Mesnil du Buisson, F. and Mauleon, P. (1972) Effect of mating on the onset and duration of ovulation in the sow. *Journal of Reproduction and Fertility* 31, 327–330.

Soede, N.M. and Kemp, B. (1993) In synchronized pigs, the duration of ovulation is not affected by insemination and is not a determinant for early embryonic diversity. *Theriogenology* 39, 1043–1053.

Soede, N.M., Helmond, F.A. and Kemp, B. (1994) Periovulatory profiles of oestradiol, LH and progesterone in relation to oestrus and embryo mortality in multiparous sows using transrectal ultrasonography to detect ovulation. *Journal of Reproduction and Fertility* 101, 633–641.

Stork, M.G. (1979) Seasonal reproductive inefficiency in large pig breeding units. *Veterinary Record* 104, 49–52.

Tanida, H., Murata, Y., Tanaka, T. and Yoshimoto, T. (1989) Mounting efficiencies, courtship behaviour and mate preference of boars under multi-sire mating. *Applied Animal Behaviour Science* 22, 245–253.

Tonetta, S.A. and diZerega, G.S. (1990) Local regulatory factors controlling folliculogenesis in pigs. *Journal of Reproduction and Fertility* Suppl. 40, 151–161.

Warnick, A.C., Casida, L.E. and Grummer, R.H. (1950) The occurrence of oestrus and ovulation in postpartum sows. *Journal of Animal Science* 9, 66–72.

Warnick, A.C., Wiggins, E.L., Casida, L.E., Grummer, R.H. and Chapman, A.B. (1951) Variations in puberty phenomena in inbred gilts. *Journal of Animal Science* 10, 479–493.

Weitze, K.F. and Waberski, D. (1996) Influence of seminal plasma on the timing of ovulation and sperm transport. *Journal of the British Fertility Society* 1(1), 14–15A.

Winfield, C.G. (1980) Oestrus detection in pigs. *Reviews in Rural Science* 4, 83–86.

Wuttke, W., Pitzel, L., Knoke, I. and Jarry, H. (1995) Interactions between TNF to induce luteolysis in porcine corpora lutea. *Biology of Reproduction* 52 (Suppl. 1), p. 64.

Artificial Control of Oestrus and Ovulation in Pigs 3

3.1. Introduction

The development of effective hormonal procedures for oestrus control would enable the pig producers to plan all aspects of their breeding programmes on a time basis, with all matings or inseminations in a particular group of pigs being concentrated into a few days. Controlled reproduction techniques have been aimed at the cyclic gilt rather than at the sow, which can be controlled to some extent by management procedures. Varley (1983a,b) observed that the need to control oestrus in groups of cyclic gilts arose from structural changes within the pig industry. Instead of a few gilts being reared each year under small units, it has become necessary to plan the rearing and systematic introduction of hundreds of gilts a year into the breeding herd. A review of measures used in the control of oestrus and ovulation in the USA was provided by Day (1989).

Although a very effective oestrus control procedure was developed and made available commercially in the UK and other countries during the 1960s, the pituitary inhibitor employed, methallibure, was subsequently withdrawn from the market. Despite this, the compound was apparently used in some Eastern European countries into the 1990s, and there is some relevance in including it in the present discussion. There were a number of changes described by Dingeldein *et al.* (1991) following the unification of Germany; these involved the replacement of the routine use of hormone treatment for oestrus and ovulation control in pigs in the former East Germany by individual checks after exposure to teaser boars.

3.1.1. Oestrus control by management procedures

In contrast to what can be done with the farm ruminants, a considerable measure of oestrus control is possible in pigs by management. In sows, the postweaning oestrus can be expected to occur 4–5 days after weaning, although this may not always be as predictable an event as has been suggested. Among postpubertal gilts, on the other hand, oestrus occurs at random throughout a 3-week period and management alone is no more likely to be effective in achieving its synchronization that it is in cattle and sheep.

3.1.2. Duration of lactation

Sows weaned after lactations of 4–8 weeks' duration can be expected to return in oestrus 3–7 days later; absence of oestrus for more than 10 days is often regarded as a reproductive problem. There is plenty of evidence to show that factors such as breed, herd, age, parity, season and nutrition all influence the interval between weaning and oestrus (Hughes and Varley, 1980; Fahmy, 1981; Martinat-Botte *et al.*, 1985; Daza *et al.*, 1989). There is general agreement that the interval progressively increases as the duration of lactation is shortened from about 6 weeks to within a week of farrowing. Daza *et al.* (1989), for example, showed that when the duration of lactation was shorter than 30 days, the interval was significantly longer compared with lactations of 30–40 days or longer. Absence of oestrus after weaning usually indicates a lack of ovarian activity, although 'silent heats' (ovulation unaccompanied by oestrus) cannot be ruled out (Stork, 1979).

3.1.3. Parity

Many authors have drawn attention to differences in the performance of primiparous and multiparous sows in their return to oestrus. In France, Martinat-Botte *et al.* (1985) recorded that, after a lactation of 3 weeks' duration or more, 24% of primiparous sows did not return in oestrus during the week after weaning (days 3 to 9) compared with 11% of the multiparous sows (Table 3.1); returning to oestrus late was associated with lower fertility rather than earlier matings. In Sweden, Tsuma *et al.* (1995) examined peripheral plasma patterns of LH, cortisol, corticosteroid-binding globulin and β-endorphin in weaning sows that resumed cyclicity and those that

Table 3.1. Incidence of oestrus after weaning and farrowing rate according to parity (from Martinat-Botte *et al.*, 1985).

	Oestrus (%)			Farrowing rate (%)		
	3–9 days	>10 days	No. of sows	3–9 days	>10 days	No. of sows
Primiparous sows	76**	24	9,557	82*	78*	10,368
Multiparous sows	89**	11	27,369	83**	73**	30,087

**P*<0.05, ***P*<0.01.

remained anoestrous after weaning. They record that basal LH concentrations were significantly higher in the sows that resumed cyclicity; corticosteroid-binding globulin and β-endorphin levels tended to be higher in sows that remained in anoestrus.

3.1.4. The effect of social environment

Many factors, such as breed, age, parity, season and nutrition are known to influence the interval between weaning and remating in sows; there is also the social environment. In Australia, Pearce and Pearce (1992) showed that exposure of weaned sows to an oestrous sow or a boar significantly increased the percentage of sows showing oestrus within 10 days of weaning; the onset of oestrus was also better synchronized than in untreated sows.

3.1.5. Effect of light environment

Sows usually show oestrus 4–8 days after weaning, but this interval is often prolonged in summer, particularly in primiparous animals; lengthening daylight in the summer may contribute towards this delay. The endocrine mechanisms underlying the effect of the light environment on the return to oestrus after weaning are not well understood. In France, Kermabon *et al.* (1995) examined the influence of daylength on gonadotrophin profiles in lactating sows. After exposure to a short daylength environment, a significantly higher proportion of sows exhibited oestrus shortly after weaning than in the long daylength group; these results confirmed the findings of Prunier *et al.* (1994) on the influence of light duration on the return to oestrus after weaning.

3.1.6. Use of AI in sows after weaning

Studies by Weitze *et al.* (1994) in Germany, using teaser boars to establish oestrus and ultrasonography to detect ovulation, led them to draw up a set of recommendations for the practical application of AI after weaning sows:

1. Sows coming into oestrus early after weaning (day 3–4) generally show oestrus for 3 days, so insemination should be started on day 2 or 3 of oestrus;
2. sows with an average interval from weaning to onset of oestrus (5 days) show oestrus for about 2 days and should be inseminated 24h after the onset of oestrus, the second insemination to be carried out 12h later;
3. sows entering oestrus late after weaning (day 6 and beyond) tend to show oestrus for only a short period (1 day) indicating that insemination should be

performed within 24h of the detection of oestrus without a second insemination being necessary.

3.1.7. Use of PMSG

A number of authors have reported on the use of gonadotrophins in the recently weaned sow in attempts to gain better control of their reproductive function. Heinze and Stahl (1989), in the former East Germany, injected 7062 sows in the January–April period with 1000 IU PMSG 24h after weaning (35 days of lactation) and recorded their performance after insemination; the conception rate was 82.5% and litter size averaged 11.6 piglets. Also in that country Rubo and Bergfeld (1990), recorded synchronizing ovulation in 2126 sows by PMSG prior to insemination at a fixed time. The advantages and disadvantages of routine synchronization of ovulation and AI were discussed by Konig and Huhn (1991), with special reference to conditions in the former East Germany, where ovulation was apparently induced in 74% of all gilts and sows. Unfortunately, in many of these reports, the applications were without controls, and details too sparse for any useful evaluation.

3.1.8. Use of anti-PMSG

In Canada, Kirkwood *et al.* (1995) reported on the effect of immunoneutralization of PMSG at a gonadotrophin-induced oestrus on the reproductive performance of sows. Sows were given an injection of 750 IU PMSG at weaning and 750 IU monoclonal anti-PMSG serum at mating. Farrowing rates were higher in the sows given PMSG followed by anti-PMSG (89%) than in controls (74%); there was no effect of treatment on subsequent litter sizes.

3.1.9. Use of oestradiol benzoate

It is well established that precocious puberty can be induced (with varying degrees of success) in the female pig by the administration of oestradiol benzoate (see Chapter 9). In Japan, Ogasa *et al.* (1994) investigated the response of sows to oestradiol benzoate, administered either on the day of weaning or 2 days later; oestrous responses and farrowing rates were influenced positively or negatively according to oestradiol dose. There appeared to be no practical merit in this approach.

3.1.10. Oestrus control in lactating sows

Although it is generally accepted that the sow does not show heat during lactation, there do appear to be some circumstances under which it is possible to overcome the inhibitory effect on ovarian activity. A system of management involving the groupings of sows, together with their litters, in the presence of a boar and feed provided *ad libitum* proved effective in getting lactating sows in oestrus about 5 weeks after parturition (Rowlinson *et al.*, 1975). This was taken as a clear example of how the pig's reproductive processes can be profoundly affected by short-term environmental factors.

3.2. Early Efforts at Oestrus Control in Pigs

Oestrus control methods have been the subject of many investigations over the years. Initially, it was often a matter of trying to get gilts showing oestrus within a day or two of one another so that a short mating period could be achieved. In later efforts, emphasis was directed towards precise control of ovulation so as to permit breeding by AI to be carried out at an assigned hour.

3.2.1. Use of progesterone and progestagens

The occurrence of oestrus and ovulation in the pig, as in cattle and other farm animals, may be controlled either by employing agents which interrupt or compounds that prolong the normal oestrous cycle. These preparations may act either by suppressing the release of the pituitary gonadotrophins and thereby delaying oestrus or by inducing premature regression of the corpora lutea; after employing such agents, follicular development and oestrous symptoms can be expected to occur 4–8 days after treatment. Progesterone or synthetic progestagens were the hormones most commonly employed to inhibit ovarian activity in early studies. Daily doses of progesterone given by injection were used to suppress oestrus and, where the dose level was adequate, this could result in acceptable fertility at withdrawal (Ulberg *et al.*, 1951; Gerrits *et al.*, 1963).

3.2.2. Use of medroxyprogesterone acetate

Several of the synthetic progestagens, when administered in early studies by oral doses or injection, were shown to be capable of inhibiting oestrus and ovarian activity; however, difficulties could arise from inadequate suppression of oestrus, reduced fertility on treatment withdrawal and the formation of cystic follicles. Medroxyprogesterone acetate (MAP) was one progestagen examined in the course of many studies in the 1950s and early 1960s; provided an appropriate dose level was used, MAP did inhibit oestrus without producing cystic follicles, but heat periods were not always well synchronized and there could be an adverse effect on litter size. The use of other synthetic progestagens, including chlormadinone (CAP), gave similar results.

These early studies showed that the pig is very sensitive to the dose of progestagen employed; however, cystic ovaries could occur if levels of progestagen treatment were either too low or too high. It was evident that progesterone and the synthetic progestagens used up to the 1970s (MAP/CAP) were not suitable agents for use in commercial practice because of the decreased fertility apparent at the synchronized heat, an unwelcome incidence of cystic follicles and a lack of precise synchronization. It was only after the withdrawal of the non-steroidal agent, methallibure, from commercial use in pigs that there was a resurgence of interest in using progestagens in oestrus control.

3.2.3. The methallibure era

Of the various farm mammals, it was in the pig that the first precise control of ovulation was achieved, using methallibure (synonyms = AIMAX or ICI33828), a compound quite unrelated to any of the steroids. Apparently, a chance observation led to the discovery that the agent would inhibit the release of pituitary gonadotrophins in the rat; experiments conducted soon afterwards showed methallibure to be highly effective in suppressing oestrus and ovulation in the pig (Polge, 1965). The compound is known to be active at the level of the hypothalamus rather than the pituitary (Kesner *et al.*, 1987). When methallibure was administered orally in daily doses of 100 mg for an 18–20-day period, follicular growth and development was effectively inhibited, although the lifespan of the corpora lutea was not affected; after treatment ended, follicular development was resumed, leading to oestrus within 5–7 days in most animals.

It quickly became apparent that fertility (farrowing rate and litter size) was not adversely affected when methallibure was used in oestrus control in gilts; unlike experiences with the progestagens, cystic follicles were not observed. Although the precise mode of action of methallibure was not understood at that time, it apparently acted on the central nervous system to induce an effective inhibition of gonadotrophin release without having any adverse effect on the pig's reproductive tract.

It was unfortunate, but essential, that the compound had to be withdrawn almost overnight in the UK and other countries because of a teratogenic effect. It was work in Canada that first showed that methallibure could result in malformations in the fetuses of sows consuming the agent in early pregnancy (King, 1969). The agent continued to be used in Eastern European countries, where it was administered under strictly regulated conditions (Jochle and Lamond, 1980).

3.2.4. Progestagens in later studies

Studies with altrenogest

Although the progestagens had been tried and generally found wanting in the premethallibure era because of adverse effects on fertility, there was a revival of interest in these steroids for oestrus control in the 1970s with the withdrawal of methallibure and the failure of prostaglandins to provide a viable option. As pig production systems became increasingly cost and labour conscious, the need to arrange reproductive events into some predetermined schedule was seen to be all the more important.

In the early 1970s, a new synthetic progestagen, altrenogest (synonyms = RU 2267, allyl trembolone, Regumate) became available and was extensively tested in pigs; by the end of that decade, results from several studies had shown evidence of effective control of oestrus and ovulation when altrenogest was fed daily over an 18-day period at a dose level of 12.5–15 mg day^{-1} per gilt (Table 3.2). Day (1984) found that the higher (20mg) dose of altrenogest is most beneficial when gilts are group-fed. Literature on the use of altrenogest has been reviewed by Webel and Day (1982), Day (1984) and Martinat-Botte *et al.* (1985).

In some of the earliest studies, it was found that although fertilization rate was not adversely affected by treatment (O'Reilly *et al.*, 1979) the incidence of cystic follicles was sufficiently high to warrant some concern (Davis *et al.*, 1979). In the light of subsequent results, however, this cystic follicle problem may have arisen from an insufficient dose of progestagen due to various environmental and animal factors.

The onset of oestrus after withdrawal of altrenogest has been reported as 4–10 days, with most heats occurring between days 5 and 7 (Table 3.3); the heat period itself appeared to be of normal duration. In studies at Cambridge,

Table 3.2. Oestrus in gilts after treatment with altrenogest (from Day, 1984).

Daily dose of altrenogest (mg)	No. treated	Percentage of gilts exhibiting oestrus	Average post-treatment interval to oestrus (days)
10.0	65	78	6.2 ± 1.3
12.5	65	97	6.9 ± 1.2
15.0	65	89	7.0 ± 1.0
12.5	25	80	4.9 ± 1.0
15.0	60	97	5.6 ± 0.8
10.0	23	82	4.3 ± 0.2
20.0	24	100	4.6 ± 0.2
40.0	24	100	6.1 ± 0.2
15.0	81	98	5.4 ± 0.1
15.0	77	100	5.3 ± 0.1

Table 3.3. Synchronization of oestrus by the use of altrenogest in gilts (from Boland and Gordon, 1981).

Dose level (mg)		Nil	15	20
No. of gilts treated		20	17	18
No. in oestrus within 10 days of progestagen withdrawal		12	16	18
Total no. in oestrus		20	17	18
Interval (days)	Mean	8.4	5.5	5.5
for all gilts	± SE	1.1	1.2	1.0

doses of 12–15 mg altrenogest were used over an 18-day period in 250 gilts (Polge, 1982); most gilts were in oestrus between 5 and 7 days after treatment and they produced litters of normal size. Unlike experiences with methallibure, the use of altrenogest was not found to depress appetite. According to Varley (1983b) toxicology studies in which altrenogest was fed to pregnant gilts did not invoke a teratogenic response and there had been no report of fetal abnormalities.

In Japan, Itoh *et al.* (1988) examined various dose levels of altrenogest (5, 10, 15 and 20 mg day^{-1}) in gilts and sows fed over an 18-day period; pigs fed at the lowest level (5 mg) failed to show oestrus. The interval from drug withdrawal to oestrus tended to increase with increasing doses over the 10–20 mg day^{-1} dose range. In later studies, Itoh *et al.* (1992) investigated oestrus synchronization in gilts fed 20 mg altrenogest from either pro-oestrus (day 19) or oestrus (day 3). For gilts that started treatment on days 19 and 3, the percentage showing oestrus synchronization was 93.8 and 95.2, respectively; there was no significant difference either in ovulation rate or fertilization rate between the two groups.

In the USA, Rhodes *et al.* (1991) examined the effect of 'flushing' and altrenogest treatment on litter size in gilts; altrenogest was fed for 15 days and flushing commenced on day 9 of treatment. They recorded significantly higher ovulation rates in flushed gilts, regardless of whether they received the progestagen. Also in the USA, the efficacy of altrenogest in synchronizing oestrus in gilts and sows in two pig-breeding programmes in Virginia was reported by Wood *et al.* (1992) who found no significant effect of treatment (15 mg day^{-1}) on litter size.

3.2.5. Synchronization with norgestomet ear implants

In some areas of research (e.g. transgenic pig programmes) it has been necessary to consider methods that do not involve feeding progestagens to control oestrus in pigs. The addition of altrenogest to feed requires pigs to be individually housed and fed, thus increasing demands for space and costs per experimental animal. In the USA, Selgrath and Ebert (1993) described the use of a norgestomet ear implant placed subcutaneously one inch from the base of the ear to synchronize group-housed pigs. The data presented suggest that the method synchronized oestrus with efficiencies equal to that reported for single-housing methodologies.

3.3. Prostaglandins in Oestrus Control

The synchronization of oestrus, using two spaced doses of $PGF_{2\alpha}$ as carried out in cattle, or even in sheep and goats, is not possible in pigs. This is because of the long refractory period of the pig's corpus luteum which means that it does not respond to $PGF_{2\alpha}$ or any of its analogues until day 12 or so of the oestrous cycle. In the 1970s, there was a suggestion that the lack of susceptibility of the corpus luteum to PG might be due to a luteotrophic effect of oestrogen. Some workers employed higher than normal doses of the PG analogue (cloprostenol) in slow-release formulations with or without an antioestrogen (Jackson and Hutchinson, 1980); such treatments, however, proved ineffective.

In the pig, as in other farm mammals, theca cells synthesize androgens, which are then converted into oestrogens under the influence of the enzyme aromatase in the granulosa cells of the ovarian follicles. Corpora lutea originate from both types of cells and it is known that these bodies are able to produce oestrogens (Przala *et al.*, 1984) and that there is an increase in oestradiol production as the luteal phase progresses (Gregoraszczuk, 1992). More recently Wuttke *et al.* (1995) have shown that during the normal process of luteolysis in the pig, the corpus luteum is invaded by macrophages which produce the cytokine TNF (tumour necrosis factor); the indications are that TNF inhibits luteal oestradiol production. It is believed that due to this inhibition, an indirect oestradiol-mediated luteotrophic effect of $PGF_{2\alpha}$ is prevented and its direct luteolytic effect becomes dominant; it is apparently necessary for both TNF and $PGF_{2\alpha}$ to act synergistically in the process of luteolysis in the pig.

3.3.1. Effect of repeated PG doses

Although of research rather than commercial interest, it is evident that the repeated administration of $PGF_{2\alpha}$ during the early luteal phase can result in premature luteolysis in the pig. In North Carolina, Estill *et al.* (1993) administered doses of 12.5mg $PGF_{2\alpha}$ (dinoprost tromethamine) every 12h on days 5–10 of the oestrous cycle; the average interval to oestrus in PG-treated gilts was significantly reduced from 19.8 to 13.3 days. The authors concluded that the gilt is susceptible to the luteolytic effects of prostaglandin before day 12 if repeated injections are given from day 5 to day 10. In attempting to explain these results, they suggested that $PGF_{2\alpha}$ may have caused up-regulation of its own receptor such that the PG receptor concentration increased earlier in the cycle, thereby rendering the corpus luteum sensitive to PG. Based on hormone profiles, it appeared that the oestrus subsequent to PG-induced premature luteolysis was normal and likely to be fertile. However, Estill *et al.* (1995) found evidence that PG-induced premature luteolysis was not mediated by an increase in the luteal PG receptor concentration; it was evident that the PG treatment resulted in premature functional and structural luteal regression.

3.3.2. Prostaglandin after prolonged oestrogen treatment

It has been known for some time that the administration of oestrogen to pigs with active corpora lutea may result in a luteotrophic rather than a luteolytic effect. Oestrogens administered towards the end of the luteal phase (days 10 to 14) may result in the persistence of the corpora lutea; such corpora lutea which are maintained well beyond their normal lifespan will regress synchronously after prostaglandin treatment (Guthrie and Polge, 1976).

Quite apart from oestrus control, it is as well to keep in mind the possible effect of oestrogenic compounds on the pig's cycle when dealing with breeding animals under some farming conditions. It is clear from several reports that if oestrogenic preparations are inadvertently ingested by gilts or sows with active corpora lutea, prolonged anoestrus may occur. For example, pigs may eat mouldy feed which contains the oestrogenic mycotoxin zearalenone; if diagnosed, sows with persistent corpora lutea arising from such mishaps can be treated effectively with $PGF_{2\alpha}$.

3.3.3. Accessory corpora lutea

In an effort to overcome the problem of lack of susceptibility of the pig's corpora lutea to prostaglandin, various novel routes to a solution were explored; in practical terms, they are of historic interest only. One approach to regulating the oestrous cycle, which is possible in pigs, but not in the other farm animals, is the induction of accessory corpora lutea in cyclic gilts and allowing these bodies to regress in the normal way. The administration of PMSG followed by hCG will induce ovulation and the formation of 'accessory' corpora lutea at any stage of the oestrous cycle; such accessory bodies regress after a normal lifespan and oestrus could be expected 18–24 days after the hCG injection. Although this approach gave some measure of control, it is far from precise because of the variability in the duration of luteal function. A refinement of the procedure uses prostaglandin when the accessory corpora lutea reach 12 days of age and become susceptible to the action of that agent.

Corpora lutea that are maintained because of pregnancy are sensitive to the lytic effects of prostaglandins. The possibility has been mentioned of prolonging the gilt's luteal phase by initiating pregnancy and then inducing regression of the corpora lutea in the early weeks of gestation; such a devious route to oestrus control is unlikely to have practical appeal.

3.4. Fixed-Time Artificial Insemination in Pigs

It was in pigs, using methallibure in conjunction with PMSG and hCG, that fixed-time AI was first shown to be a highly effective means of achieving pregnancy in farm animals without reference to oestrus (Polge *et al.*, 1968). After receiving 100 mg of methallibure daily for 20 days, gilts were given PMSG 24 h after that treatment and hCG after a further 4 days; work in Cambridge showed a 90% conception rate and normal litter sizes in gilts inseminated 24h after hCG without reference to oestrus. It seemed, at that time, that the methallibure–PMSG–hCG regimen had removed a major obstacle to the commercial exploitation of AI in breeding gilts. The withdrawal of methallibure from the UK market was a serious blow to hopes of developing this area of AI application in pigs.

In Eastern Europe, however, experimental work continued with methallibure and was still being reported in the 1990s. One modification of the fixed-time AI treatment was to include GnRH rather than hCG in the treatment regimen. In the former East Germany, Huhn *et al.* (1991) reported studies in which gilts on ten farms were treated orally with methallibure for 15 days; on day 15, PMSG (100IU) was injected at 8a.m., followed by 500IU hCG or 50µg GnRH at 3p.m. After both treatments, gilts started to ovulate before 6a.m. on day 16. The pigs were inseminated 24 and 42h after hCG or GnRH treatment resulting in a pregnancy rate in GnRH-treated gilts 4.4% higher than in those treated with hCG.

Dingeldein *et al.* (1991) give an account of post-unification changes in the reproductive management of sows and gilts at large pig-breeding units in the former East Germany. Such changes involved the replacement of the routine use of hormone treatment for the synchronization of oestrus and ovulation by individual oestrus checks after exposure to teaser boars.

3.4.1. Artificial insemination after altrenogest treatment

In France, from 1983, altrenogest was extensively tested in gilts in different environments and under various management systems (Martinat-Botte *et al.*, 1985). The close synchronization obtained with altrenogest treatment allowed the use of fixed-time artificial insemination on days 6 and 7 after the last progestagen feeding and without the detection of oestrus; the time of insemination had to be adjusted according to the genotype of gilts. In the former East Germany, Busch *et al.* (1988) recorded experiences in using altrenogest in combination with PMSG and GnRH/hCG in which AI was carried out at a fixed time (22 and 28–40h after completion of treatment).

A further report (Busch *et al.*, 1992) recorded data from four trials in which oestrus was synchronized in 2001 gilts by treatment with 20mg altrenogest daily for 8–18 days, with or without 300–900IU PMSG administered 24h after the last progestagen dose, followed by treatment with 300IU hCG and 300IU GnRH given 76–78h after PMSG injection. There was no significant difference in conception rate or litter size between gilts treated with altrenogest for 15 days followed by 500IU PMSG + 300IU hCG + 300IU GnRH and gilts treated with methallibure in association with PMSG/hCG/GnRH. In Korea, normal fertilization and development of embryos in gilts treated with altrenogest (18 days at 20mg day^{-1}) and gonadotrophins was reported by Lee *et al.* (1992a,b). They adminstered PMSG and hCG on days 1 and 4, respectively, after the withdrawal of the progestagen.

3.4.2. AI after detection of oestrus

Martinat-Botte *et al.* (1990) refer to a preliminary survey of field use which suggested that altrenogest treatment gave not only good fertility, but also increased litter size. This is when breeding is carried out after oestrus has been detected in the synchronized gilts.

In their experimental work, they used altrenogest over an 18-day period (20mg day^{-1}); inseminations (6000 million sperm per dose) were carried out at 24h intervals, the first at 6–12h after the onset of oestrus. The results of their study are shown in Table 3.4. According to Marinat-Botte *et al.* (1990), these results supported field observations in France and confirmed data recorded elsewhere (Varley, 1983a,b; Johnston, 1988). They concluded that under conditions of practical pig production, an improved farrowing rate, together with an increase in litter size, could be expected from pubertal gilts treated with altrenogest. Ashworth *et al.* (1992), who used 20 mg daily doses of altrenogest over 18 days to Meishan and Landrace × Large White gilts, recorded a non-significant increase in the ovulation rate of both breed types. Martinat-Botte *et al.* (1995) recorded that 93% of gilts treated with altrenogest (18 days at 20 mg day^{-1}) were in oestrus by 7 days after progestagen withdrawal; ovulation rate was increased by treatment (15.4 vs. 14.6).

Table 3.4. Farrowing rate and litter size after treatment with altrenogest (from Martinat-Botte *et al.*, 1990).

	Control		Altrenogest
Number of gilts	281		242
Farrowing rate	80.8%	$P<0.05$	88.4%
Piglets born (total)[a]	9.1 ± 2.3	$P<0.05$	9.6 ± 2.3
Piglets born alive[a]	8.5 ± 2.3	$P<0.02$	9.1 ± 2.4
Birth weight (kg)[a]	1.6 ± 0.2	ns	1.6 ± 0.2

ns, Not significant; [a]Mean ± SD.

3.4.3. Controlling time of ovulation

Attempts have been made to influence reproductive performance in pigs by gaining greater control over the occurrence of ovulation, relative to the onset of oestrus. In Germany, Heinze *et al.* (1991) reported on the use of a GnRH analogue (Gonavet), at a dose level of 25 or 50µg, in 10,800 sows for the synchronization of ovulation. In the Netherlands, Meulen *et al.* (1993) used GnRH (100µg GnRH analogue) at the onset of standing oestrus and slaughtered gilts at various intervals after breeding (5–35 days); although treated animals had more corpora lutea (16.4 vs. 14.3) and embryos (15.4 vs. 13.0) the differences were not significant.

3.5. Inducing Ovarian Activity in Acyclic Anovulatory Pigs

3.5.1. Nutritional anoestrus in young sows

Anoestrus is one of the most common reproductive problems of sows and gilts and in particular among sows that have been weaned from their first litter. King and Williams (1981) recorded that about 50% of first litter sows in their study failed to exhibit oestrus within a week of weaning whereas almost 80% of older sows were mated within this period. King (1987) examined available information on the the relationship between nutritional factors and postweaning anoestrus in young pigs.

3.5.2. Alleviating the problem of anoestrus

Evidence from studies in the USA makes it clear that season of the year affects the duration of the postweaning anoestrus in sows (Britt *et al.*, 1983). During summer, postweaning anoestrus, particularly in primiparous sows, may be related to daylength. Energy intake during lactation may decrease in summer due to reduced feed consumption under high ambient temperature conditions. Various methods have been used in attempts to alleviate the problem of seasonal anoestrus; these include increasing energy intake, providing an artificially cooled environment, manipulating the light environment and using hormone treatments.

Dial and BeVier (1986) found that PMSG has been used therapeutically in many European countries to induce a fertile oestrus in persistently anoestrous sows. Other gonadotrophins are not as effective. Although hCG will induce ovulation, it is relatively ineffective in initiating oestrus. Combinations of PMSG (400IU) and hCG (200IU) have been effective in treating pigs showing postweaning anoestrus.

Studies reported by Britt *et al.* (1986) demonstrated that a 1200IU dose of PMSG given on the day after weaning was effective in promoting prompt return to oestrus in weaned primiparous sows, with 95% of the animals being detected in oestrus within 10 days in contrast to 58% of control sows. Feeding altrenogest for 3 days before giving the PMSG was not regarded as necessary.

3.6. References

Ashworth, C.J., Haley, C.S. and Wilmut, I. (1992) Effect of Regumate on ovulation rate, embryo survival and conceptus growth in Meishan and Landrace × Large White gilts. *Theriogenology* 37, 433–443.

Boland, M.P. and Gordon, I. (1981) Attempts to control the oestrous cycle in the gilt. *Irish Veterinary Journal* 35, 82–85.

Britt, J.H., Szarek, V.E. and Levis, D.G. (1983) Characterization of summer infertility of sows in large confinement units. *Theriogenology* 20, 133–140.

Britt, J.H., Esbenshade, K.L. and Heller, K. (1986) Responses of seasonally anestrous gilts and weaned primiparous sows to treatment with pregnant mare's serum gonadotropin and altrenogest. *Theriogenology* 26, 697–707.

Busch, W., Maab, P., Freistedt, R. and Wohlfahrt, E. (1988) The application of allyl tenbolone (Regumate) for synchronization of oestrus in gilts under big farm conditions. *Proceedings of the 11th International Congress of Animal Reproduction and AI* (Dublin), Paper 465.

Busch, W., Poirier, P., Maass, P., Wohlfahrt, E. and Freistedt, R. (1992) Investigations on oestrus synchronization in sows by means of altrenogest (Regumate). *Monatshefte fur Veterinarmedizin* 47, 307–316.

Davis, D.L., Knight, J.W., Killian, D.B. and Day, B.N. (1979) Control of estrus in gilts with a progestagen. *Journal of Animal Science* 49, 1505–1509.

Day, B.N. (1984) Estrous cycle regulation. *Proceedings of the 10th International Congress Animal Reproduction and AI* (Urbana) IV, 1–8.

Day, B.N. (1989) Control of oestrus and ovulation in swine. *Agri-Practice* 10, 11–15.

Daza, A., Ovejero, I., Perez-Guzman, M.D. and Buxade, C. (1989) Analysis of factors that influence the length of the weaning-to-conception interval of the sow. *Investigacion Agrania, Produccion y Sanidad Animlaes* 4, 163–173.

Dial, G.D. and BeVier, G.W. (1986) Pharmacologic control of estrus and ovulation in the pig. In: Morrow, D.A. (ed.) *Current Therapy in Theriogenology.* W.B. Saunders, Philadelphia, pp. 912–914.

Dingeldein, J., Giesen, H., Klingelholler, A., Holtz, W., Huhn, E.U. and Rickert, J. (1991) Production of piglets without hormones. *Tierzuchter* 43, 294–296.

Estill, C.T., Britt, J.H. and Gadsby, J.E. (1993) Repeated administration of prostaglandin F2 α during the early luteal phase causes premature luteolysis in the pig. *Biology of Reproduction* 49, 181–185.

Estill, C.T., Britt, J.H. and Gadsby, J.E. (1995) Does increased PGF2-alpha receptor concentratioin mediate PGF2-alpha-induced luteolysis during early diestrus in the pig? *Prostaglandins* 49, 255–267.

Fahmy, M.H. (1981) Factors influencing the weaning to oestrus interval in swine: a review. *World Review of Animal Production* 17, 15–27.

Gerrits, R.J., Fahning, M.L., Meade, R.J. and Graham, E.F. (1963) The effect of synchronization of estrus on fertility in gilts. *Journal of Animal Science* 21, 1022–1025.

Gregoraszczuk, E.L. (1992) Interrelationship between steroid hormone secretion and morphological changes of porcine corpora lutea at various periods of luteal phase. *Endocrine Regulators* 26, 189–194.

Guthrie, H.D. and Polge, C. (1976) Luteal function and oestrus in gilts treated with a synthetic analogue of prostaglandin F2-alpha (ICI 79939) at various times during the oestrous cycle. *Journal of Reproduction and Fertility* 48, 423–425.

Heinze, A. and Stahl, U. (1989) Investigation on the stimulation of oestrous cycle in the summer months in sows with synchronized ovulation, using different doses of PMSG. *Monatshefte fur Veterinarmedizin* 44, 569–570.

Heinze, A., Stahl, U. and Raasch, M.L. (1991) Synchronization of ovulation in sows. Investigations on the dose of Gonavet in relation to month of mating. *Monatshefte fur Veterinarmedizin* 46, 210–212.

Hughes, P. and Varley, M. (1980) *Reproduction in the Pig*. Butterworths, London, 241 pp.

Huhn, U., Raasch, M.L. and Konig, I. (1991) Pregnancy rate in gilts after oestrus synchronization and fixed-time insemination; results of one-year field experiments. *Reproduction in Domestic Animals* 26, 126–135.

Itoh, Y., Kondo, Y., Narashima, T., Yanaka, T. and Masuda, H. (1988) Control of oestrus and level of serum progesterone in pigs after oral administration of altrenogest. *Japanese Journal of Swine Science* 25, 133–141.

Itoh, Y., Kondo, Y., Uchiyama, K. and Masuda, H. (1992) Oestrous synchronization in pigs by oral administration of altrenogest from pro-oestrus or oestrus. *Japanese Journal of Swine Science* 29, 152–157.

Jackson, P.S. and Hutchinson, F.G. (1980) Slow release formulations of prostaglandin and luteolysis in the pig. *Veterinary Record* 106, 33–34.

Jochle, W. and Lamond, D.R. (1980) *Control of Reproductive Functions in Domestic Animals*. Gustav Fischer Verlag, Jena, 248 pp.

Johnston, N.E. (1988) The benefits of Altrenogest in synchronizing oestrus in gilts. *Proceedings of the 10th International Congress of Pig Veterinary Society* (Rio de Janeiro), p. 281.

Kermabon, A.Y., Prunier, A., Djiane, J. and Salesse, R. (1995) Gonadotropins in lactating sows exposed to long or short days during pregnancy and lactation: serum concentrations and ovarian receptors. *Biology of Reproduction* 53, 1095–1102.

Kesner, J.S., Kraeling, R.R., Rampacek, G.B. and Johnson, B. (1987) Absence of an estradiol-induced surge of luteinizing hormone in pigs receiving unvarying pulsatile gonadotropin-releasing hormone stimulation. *Endocrinology* 121, 1862–1869.

King, G.J. (1969) Deformities in piglets following administration of methallibure during specific stages of gestation. *Journal of Reproduction and Fertility* 20, 551–553.

King, R.H. (1987) Nutritional anoestrus in young sows. *Pig News and Information* 8(1), 15–22.

King, R.H. and Williams, I.H. (1981) Nutrition and reproductive performance of sows. *Proceedings of the 4th Australian Poultry and Stockfeed Convention* (Perth), pp. 93–103.

Kirkwood, R.N., Soede, N.M., Dyck, G.W. and Thacker, P.A. (1995) The effect of immunoneutralization of PMSG at a gonadotropin-induced oestrus on the duration of ovulation and reproductive performance of sows. *Animal Science* 61, 321–324.

Konig, I. and Huhn, U. (1991) Control of reproduction in sows by biotechnological methods. *Tierzucht* 45, 390–391.

Lee, S.H., Byun, T.H., Lee, J.H., Kim, H.J. and Hong, K.C. (1992b) Normal fertilization and development of early embryos in gilts treated with allyl trenbolone

(Regumate). *Korean Journal of Animal Sciences* 34, 257–262.

Lee, S.H., Hong, K.C. and Jeon, J.T. (1992a) Induction of oestrus synchronization in gilts by allyl trenbolone. *Korean Journal of Animal Sciences* 34, 253–256.

Martinat-Botte, F., Bariteau, F., Badouard, B. and Terqui, M. (1985) Control of pig reproduction in a breeding programme. *Journal of Reproduction and Fertility, Suppl.* 33, 211–228.

Martinat-Botte, F., Bariteau, F., Forgerit, Y., Macar, C., Moreau, A., Terqui, M. and Signoret, J.P. (1990) Control of oestrus in gilts. II. Synchronization of oestrus with a progestagen, Altrenogest (Regumate): effect on fertility and litter size. *Animal Reproduction Science* 22, 227–233.

Martinat-Botte, F., Bariteau, F., Forgerit, Y., Macar, C., Poirier, P. and Terqui, M. (1995) Synchronization of oestrus in gilts with altrenogest: effects on ovulation rate and foetal survival. *Animal Reproduction Science* 39, 267–274.

Meulen, J.Van Der, Helmond, F.A., Oudenaarden, C.P.J., Lende, T. Van Der and Kronnie, F.Te (1993) Effect of injection of GnRH analogue at the onset of oestrus on LH and FSH release and embryonic mortality in gilts. *Netherlands Journal Agricultural Science* 41, 37–46.

Ogasa, A., Tsutsui, T. and Kawakami, E. (1994) Response of sows to estradiol benzoate after weaning. *Journal of Veterinary Medical Science* 56, 995–996.

O'Reilly, P.H., McCormack, R., O'Mahony, K. and Murphy, C. (1979) Estrus synchronization and fertility in gilts using a synthetic progestagen (Allyl trenbolone) and inseminated with fresh stored or frozen semen. *Theriogenology* 12, 131–137.

Pearce, G.P. and Pearce, A.N. (1992) Contact with a sow in oestrus or a mature boar stimulates the onset of oestrus in weaned sows. *Veterinary Record* 130, 5–9.

Polge, C. (1965) Effective synchronization of oestrus in pigs after treatment with ICI compound 33, 828. *Veterinary Record* 77, 232–236.

Polge, C. (1982) Embryo transplantation and preservation. In: Cole, D.J.A. and Foxcroft, G.R. (eds) *Control of Pig Reproduction.* Butterworths, London, pp. 277–291.

Polge, C., Day, B.N. and Groves, T.W. (1968) Synchronization of ovulation and artificial insemination in pigs. *Veterinary Record* 83, 136–142.

Prunier, A., Dourmad, J.Y. and Etienne, M. (1994) Effect of light regimen under various ambient temperatures on sow and litter performance. *Journal of Animal Science* 72, 1461–1466.

Przala, J., Wiesak, T., Grazul, A. and Cieplinska, E. (1984) The effect of prolactin on estradiol-17β and testosterone plus 5-dihydrotestosterone secretion by porcine luteal cells in vitro. *Clinical Endocrinology* 83, 334–343.

Rhodes, M.T., Davis, D.L. and Stevenson, J.S. (1991) Flushing and altrenogest affect litter traits in gilts. *Journal of Animal Science* 69, 34–40.

Rowlinson, P., Boughton, H.G. and Bryant, M.J. (1975) Mating of sows during lactation: observations from a commercial unit. *Animal Production* 21, 233–241.

Rubo, B. and Bergfeld, J. (1990) Investigations on the accuracy of PMSG dosage for the synchronization of ovulation in gilts and sows. 3. Conception rate and litter size. *Archiv fur Experimentelle Veterinarmedizin* 44, 803–812.

Selgrath, J.P. and Ebert, K.M. (1993) Synchronization of swine with norgestomet implants. *Theriogenology* 39, 306.

Stork, M.G. (1979) Seasonal reproductive inefficiency in large pig breeding units in Britain. *Veterinary Record* 104, 49–52.

Tsuma, V.T., Einarsson, S., Madej, A. and Lundeheim, N. (1995) Hormone profiles around weaning in cyclic and anoestrous sows. *Journal of Veterinary Medicine, Series A* 42, 153–163.

Ulberg, L.C., Grummer, R.H. and Casida, L.E. (1951) The effects of progesterone upon ovarian function in gilts. *Journal of Animal Science* 10, 665–671.

Varley, M.A. (1983a) Synchronization of oestrus in gilts. *Pig News and Information* 4(2), 151–156.

Varley, M.A. (1983b) The regulation of the oestrous cycles in groups of post-pubertal female pigs using allyl trenbolone. *Animal Production* 36, 211–215.

Webel, S.K. and Day, B.N. (1982) The control of ovulation. In: Cole, D.J.A. and Foxcroft, G.R. (eds) *Control of Pig Reproduction.* Butterworths, London, pp. 197–210.

Weitze, K.F., Wagner-Rietschel, H., Waberski, D., Richter, L. and Krieter, J. (1994) The onset of heat after weaning, heat duration, and ovulation as major factors in AI timing in sows. *Reproduction in Domestic Animals* 29, 433–443.

Wood, C.M., Kornegay, E.T. and Shipley, C.F. (1992) Efficacy of altrenogest in synchronizing estrus in two swine breeding programs and effects on subsequent reproductive performance of sows. *Journal of Animal Science* 70, 1357–1364.

Wuttke, W., Pitzel, L., Knoke, I. and Jarry, H. (1995) Interactions between TNF to induce luteolysis in porcine corpora lutea. *Biology of Reproduction* 52 (Suppl. 1), p. 64.

4 Pregnancy Testing in Pigs

4.1. Introduction

Even in the most efficient of breeding herds, about 10% of services may turn out to be unsuccessful; the early and accurate detection of non-pregnant sows would be really useful to the commercial pig producer. Traditionally, sows and gilts that do not return to oestrus 18–26 days after service are presumed to be pregnant; detection of heat is carried out at that time by the stockperson with or without the aid of a boar and perhaps again at 6 weeks after breeding. As mentioned in earlier chapters, the tendency is increasingly towards pig-breeding herds becoming larger; as a consequence of this and various other factors, there is the need for an accurate and practicable method of pregnancy diagnosis.

Routine pregnancy testing is likely to be most applicable in large herds, in which there is a high barren cull rate, on farms with high in-pig gilt sales and under conditions in which artificial insemination is practised. Whatever the reasons, the diagnostic technique needs to have a high degree of accuracy in distinguishing between pregnant and non-pregnant pigs and should be capable of being applied as early as possible in pregnancy.

4.1.1. Returns-to-service

The traditional method available to the stockperson which permits the diagnosis to be made in time to permit rebreeding at 3 weeks after mating is the detection of the repeat heat. Detection of oestrus with a boar, the most widely used diagnosis of pregnancy, has been suggested as the best method, although variable responses among pig herds have been recorded (Bosc *et al.*, 1977). There is also the matter of the time and labour required in testing each sow at the appropriate time after mating. Checking for returns to service is not useful for routine testing outside the 18–24-day period after breeding, or multiples of that period. As noted by Buddle (1986), false-positives may arise with barren sows (anoestrous though not pregnant) or because of the inability of non-pregnant sows to show sexual receptivity because of various environmental or animal (e.g. bullying) factors.

Methods used over the years to diagnose pregnancy in the pig include radiography (Walker, 1972), the measurement of urinary oestrogens (Grunsell and Robertson, 1953), examination of vaginal or cervical mucus (Betteridge and Raeside, 1962) and the vaginal biopsy technique (Done and Heard, 1968; Diehl and Day, 1973; Williamson and Hennessy, 1975). According to Meredith (1990), discussing the relative merits of the different methods available, the most accurate results could be achieved by real-time ultrasound, rectal palpation and the oestrone sulphate assay.

4.2. Physiology and Endocrinology of Early Pregnancy in the Pig

4.2.1. Early days in the growth of the pig conceptus

Fertilization in the pig, as in cattle and sheep, takes place in the oviduct, at the ampullary–isthmic junction. Early development during the cleavage stages in the pig has been much less well documented than in cattle or sheep. Hancock (1961) estimated the first cleavage division to occur 20h after the initial fertilization events. According to Hunter (1974), the 2-cell stage is first found at 17–19h after ovulation, but lasts for only 6–8h; the 4-cell stage is of longer duration, 20–24h. The same author notes that pig embryos are usually at the 4-cell stage when they pass into the uterus, some 46–48h after ovulation. It is believed that activation of the embryonic genome occurs at the 4-cell stage in pigs (Jarrell *et al.*, 1991; Kronnie and de Boer, 1993).

The 8–16-cell early morula stage is reached around day 4; the first signs of compaction have been recorded at day 4, at the 8-cell stage (Hunter, 1974); the first junctional complexes are found at the compacted morula stage around day 5. The blastocyst stage is reached at day 5–6 and the cell number at the time of blastocoele formation has been given as 16–32 (Papaioannou and Ebert, 1988). Before hatching from the zona pellucida, which occurs at day 6–7, the pig blastocyst has been reported to consist of some 65–120 cells; there are those who believe that the hatching process is probably independent of cell number (Niemann *et al.*, 1983).

From day 11–12, pig blastocysts start elongating; there is a decrease in their diameter, which gives rise to filamentous structures which may be up to 100cm in length. Such elongation does not occur in all blastocysts at the same time and this gives rise to a conceptus containing blastocysts at different developmental stages. Using immunocytochemistry, King and Ackerley (1985) demonstrated the presence of oestrogen in the cells of the embryo from day 10.

4.2.2. Migration and spacing of embryos

Migration of pig embryos from the tip of the uterine horn towards the body of the uterus starts after they have hatched and during the period when the blastocyst has started to expand (Dhindsa *et al.*, 1967). Oestrogens apparently play an important part in the intrauterine migration of embryos in the pig; studies by Pope *et al.* (1982) demonstrated that silastic beads were only transported through the pig's uterus if they were impregnated with oestradiol. Oestrogen secretion can be measured in blastocysts from 5–7 mm onwards and increases markedly during expansion and early elongation (Perry *et al.*, 1976).

It appears that embryo migration is a result of myometrial activity; there are indications that oestrogens may influence the local release of histamine in the uterus resulting in an increased blood flow and greater contractility of the myometrium. Embryos migrate throughout the uterus and at the time of attachment come to be located more or less equidistant from one another. If spacing of embryos is uneven or if an excess of embryos results in uterine crowding, some embryos may not have sufficient space for optimal prenatal development (Dziuk, 1985). In general, however, and as shown in studies by Knight and Tarraf (1994) there is only a limited effect of intrauterine position on conceptus development and on placental and endometrial steroidogenic activity in the pregnant pig.

4.2.3. Attachment of embryos

Attachment starts around day 13–14 in the pig (Dantzer, 1985). The process begins with loose contact between trophoblast and uterine membranes near the embryonic disc and is completed by intermingling of uterine and trophoblastic microvilli after day 18. It is believed that embryonic oestrogens play a role in regulating the uterine secretions on which the embryos are highly dependent, not only prior to attachment, but for a period after attachment. Vallet *et al.* (1994), for example, has indicated that the pig conceptus stimulates endometrial protein secretion starting between days 40 and 60 of pregnancy. These workers suggested that identification of conceptus factors which stimulate endometrial function could be useful in increasing litter size in the pig.

The influence of sex on the early growth of pig embryos was examined by Cassar *et al.* (1994); they found that most male fetuses grew faster than female ones before the commencement of attachment to the uterus.

4.2.4. Formation of the allanto-chorion

The outermost embryonic membrane of the pig is the allanto-chorion. The chorion is formed by elongation of the trophoblast; the allantois develops as an outgrowth of the hind end of the gut and rapidly elongates and closely invests the chorionic sac. The allantois is vascularized and brings its blood vessels in close apposition with the chorion, thereby forming the vascular allanto-chorion; fusion is complete by day 24 in the pig. Formation of the allanto-chorion is of interest in terms of explaining the different ways in which the farm species are adapted to multiple pregnancies.

In the pig, there is usually no fusion of adjacent chorionic sacs and hence no likelihood of anastomosis of blood vessels of adjacent embryos (Fig. 4.1). In sheep and cattle, on the other hand, adjacent chorions fuse very early in embryonic life; the allantois of each embryo then elongates within the common chorion and, in the cow, they usually fuse where they meet. Fusion of adjacent allantoida does not occur in the sheep; in the pig, where there is normally neither fusion of chorion nor allantois, the opportunity for anastomosis of embryonic blood vessels should not occur. For such reasons, the occurrence of freemartins in the pig is most unlikely.

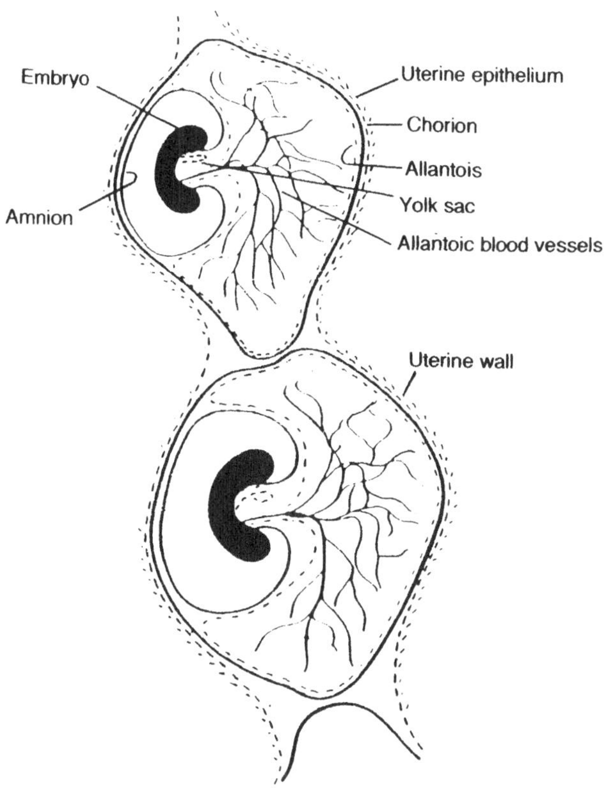

Fig. 4.1. Diagrammatic representation of the arrangement of the fetal membranes in the pig (from Robinson, 1957).

4.2.5. Maternal recognition of pregnancy in the pig

It is well accepted, from embryo transfer experiments and other studies, that the developmental stages of the uterus and the pig blastocyst must be in synchrony for successful attachment and survival of embryos. It is believed that there is a signal that is responsible for achieving the required synchrony between uterus and conceptus. The environment provided by the sow's uterus during early pregnancy is believed to have limited receptivity to the pre-attachment embryo, which must make its presence felt immediately if it is to survive; this is the process known as the 'maternal recognition of pregnancy', which is believed to be mediated by the production of oestrogens by the elongating blastocyst which results in the rapid release of endometrial secretions into the lumen of the uterus (Geisert *et al.*, 1990). Uterine secretions provide a uterine environment essential to normal development of the conceptus (Roberts and Bazer, 1988; Simmen and Simmen, 1990).

According to Bazer *et al.* (1994), in a review of pregnancy recognition in pigs, it is believed that the uterine endometrium secretes the luteolysin, PGF, and the embryos secrete oestrogens which are antiluteolytic. The PGF is secreted by the pregnant pig into the uterine lumen, where it is sequestered to exert its biological effects *in utero* and/or be metabolized to prevent luteolysis.

Maternal recognition of pregnancy can be viewed as the process by which the pig conceptus prolongs the functional lifespan of the corpora lutea established at the time of ovulation. In the pig, it has been proposed that oestrogen prevents luteolysis through reorientation of endometrial prostaglandin release, the PG being directed into the uterine lumen rather than into the uterine vasculature (Geisert *et al.*, 1990); there are indications that oestrogen may interact with prolactin and/or conceptus secretory proteins to bring about this change in direction of prostaglandin movement. Studies reported by Christenson and Ford (1995) have indicated that porcine embryos as early as day 13 contain factors which can directly prevent luteal regression at the corpus luteum level.

Maternal recognition of pregnancy in cattle and sheep is known to be initiated by the secretion of proteins which have been identified as members of the interferon-alpha (IFN-α) family. Studies have reported that porcine conceptuses secrete interferon during the pre-attachment period in early pregnancy (Cross and Roberts, 1989); it was concluded that production of IFN by the elongating porcine blastocyst is not restricted to ruminant species and may represent a more general phenomenon. In France, Bonnardiere *et al.* (1991) reported the production of two species of interferon, IFN-α and IFN-γ by pig embryos during the peri-attachment period. It is also apparent that the endometrium of the pig produces large quantities of oxytocin at the time of maternal recognition of pregnancy and that this production appears to be locally dependent on the presence of elongated blastocysts (Trout *et al.*, 1995); the implication is that endometrial oxytocin is involved in the maternal recognition of pregnancy in this species.

4.2.6. Endocrinology of early pregnancy

During pregnancy in the sow, which normally has a duration averaging 114–116 days (usual range 110–120 days), cyclic ovarian activity and ovulation are normally suppressed; for this reason, as in sheep and cattle, it is possible to determine the number of oocytes shed by examining the ovaries at any stage of pregnancy. The appearance of the ovary in pregnancy is similar to that of mid-cycle. There may be minor follicular growth and atresia taking place; this is more evident during the first half of the gestation period. On rare occasions, oestrus itself may be observed in the pregnant sow, as has been reported in cattle and sheep.

Plasma hormone levels in pregnancy in the pig have been measured by many workers using radioimmunoassay and other techniques. Progesterone concentration declines quite rapidly from its mid-cycle level to about 10–15 ng ml^{-1} at 3–4 weeks of pregnancy; the level then remains fairly constant for the remainder of the gestation period until a drop occurs shortly before the time of farrowing. Little or no effect has been evident on progesterone concentration in maternal plasma during early pregnancy from the number of embryos present in the uterus (Dziuk, 1991).

The effect of pregnancy on the uterine secretion of $PGF_{2\alpha}$ in response to oxytocin was examined by Edgerton *et al.* (1995). Their results demonstrated that uterine secretory responsiveness to oxytocin is suppressed during early pregnancy and that this effect may be mediated through oestrogen secreted by the conceptus.

Although small amounts of progesterone may be derived from the placenta in late pregnancy, most is produced by the corpora lutea; the number of fetuses that survive in the uterus does not appear to affect the level of progesterone. Studies have shown the plasma concentration remaining within the normal range and corpora lutea continuing to be functional even with all fetuses dead (Torres and First, 1975). The minimum level of plasma progesterone for the maintenance of pregnancy is about 6 ng ml^{-1} (Ellicott and Dziuk, 1973); at lower concentrations the pregnancy is lost and higher levels do not apparently increase embryo survival rate. It would seem that the level of progesterone required for the maintenance of pregnancy is an all-or-none phenomenon with little quantitative effect from various levels.

4.2.7. Maintenance of pregnancy

The maintenance of pregnancy in the sow appears to be totally dependent on the presence of functional corpora lutea; removal of the ovaries, or any factor which results in luteal failure and cessation of progesterone production, will terminate pregnancy within 24–48 h (Diehl and Day, 1974). The removal of the pituitary results in abortion in the pig, indicating that, in contrast to the corpus luteum of the oestrous cycle, that of pregnancy is presumably dependent on the continuous production of a pituitary luteotrophin. The most likely candidate to fill this role appeared to be LH; not only was treatment with anti-LH serum found to cause luteal regression and termination of pregnancy but several LH-like preparations, including LH and hCG, have been found capable of maintaining the corpora lutea of the hysterectomized/hypophysectomized sow.

Although it was initially thought that at least four corpora lutea were required to maintain pregnancy in the sow (Martin *et al.*, 1977), later studies showed that pregnancy could be maintained with as few as one or two corpora lutea, provided the reduction was done in a stepwise manner (Dziuk, 1991). There is a relationship between the plasma progesterone level and the number of corpora lutea, but no clear correlation has been shown between progesterone concentration and the number of embryos in the litter. However, it is possible that progesterone production by the corpora lutea may vary according to various factors and on occasions may be less than sufficient to provide for optimal embryo survival (Staigmiller, 1974).

4.2.8. Follicular development and uterine changes during early pregnancy

Ovarian follicular development during early pregnancy in the pig was examined by Wiesak *et al.* (1992) in Nottingham; their results showed that the growth of follicles during early pregnancy was different from that during the luteal phase of the oestrous cycle. Studies in Ireland by Ryan *et al.* (1994) used ultrasound to map ovarian follicles and to look for evidence of follicular waves during early pregnancy in the pig; they concluded that follicle waves did not occur. Anderson *et al.* (1994) examined oestrogen-associated changes in uterine blood flow and maternal recognition of pregnancy in Meishan pigs; they recorded evidence supporting the view that oestrogen secretion by the expanding pig blastocysts increases uterine blood flow and the onset of maternal recognition.

4.2.9. Daylength and seasonal effects

One factor suggested as influencing luteal function and progesterone production in pregnant pigs was daylight. Russian workers observed that the corpora lutea of sows exposed to additional light (up to 17h daily) during the winter months were heavier and apparently more active than those of sows receiving no extra light (Klotchkov *et al.*, 1971); litter size in early pregnancy and at term was also improved by the provision of additional light in winter. There were also reports around that time that litter size in sows kept in darkness was smaller than that in sows provided with a natural daylight pattern (Steger *et al.*, 1971).

Seasonal effects on reproductive function in pigs have been noted by a number of authors and certain of these effects could be associated with the duration and intensity of light (Stork, 1979; Tomes and Nielsen, 1979). In the wild pig, the evidence suggests that reproductive activity is at its maximum in the autumn and winter months (Signoret, 1980).

Basal levels of plasma LH during pregnancy have been reported to be about 1–2ng ml^{-1} (Guthrie *et al.*, 1972; Parvizi *et al.*, 1976), about double those found during the oestrous cycle (excluding the preovulatory LH surge); minor peaks of LH were evident at intervals of a few hours during pregnancy and were followed closely by increased plasma progesterone concentrations, which supported the view that production of the steroid was dependent on LH.

4.3. The Use of Ultrasonics in Pregancy Testing

In the 1960s, ultrasonic techniques were used in the diagnosis of pregnancy in pigs. First there was the Doppler instrument, followed a few years later by A-mode sound equipment; finally, there was real-time ultrasonics.

4.3.1. Doppler method

Before reaching the farm, the ultrasonic Doppler technique had been widely used in human medicine for some years and was held to be a very accurate and safe method of providing information about pregnancy. In the pig, Doppler ultrasound was used to detect the the motion of blood in the fetal heart or in the large umbilical blood vessels associated with the pig fetuses; in this, ultrasound waves emanating from the instrument's transducer strike moving blood particles and are reflected back at a slightly different frequency which can be converted into audio or visual signals. A diagnostic procedure based on Doppler signals reflected from the uterine blood vessels of the sow through an intrarectal probe was also reported in the late 1960s. Blood flow in the uterine artery could often be heard as early as 21 days after breeding but was not regarded as reliable until 30 days. In carrying out the examination, the sow was restrained in a crate; the animal was not fed at this time to avoid sounds from digestion interfering with the interpretation of signals from the uterus.

Before the advent of ultrasonic equipment, it was almost impossible to obtain information on fetal heart activity, especially in the early stages of pregnancy. Using the method, some workers reported a highly significant negative correlation between fetal heart rate and fetal age; according to this, when service dates were uncertain, the time of farrowing could be predicted with some accuracy from fetal heart rate data. Although unlikly to be of interest to the busy pig producer, there might be some occasions when such information is useful.

4.3.2. Amplitude–depth or pulse–echo devices

A few years after Doppler sound was first introduced, the amplitude–depth ultrasound procedure (also known as A-mode, A-scope, A-scan) arrived on the farm (Fig. 4.2); this technique proved to be simpler and more reliable. The method detected differences in the acoustical impedance between the contents of the pregnant uterus and the other abdominal viscera; allantoic and amniotic fluids played a major role in reflecting the echo pattern associated with early pregnancy in pigs. Work in other mammals had also shown that allantoic and amniotic fluids were not only good transmitters of ultrasound but also provided excellent contrast in accoustical impedance to the adjacent fetal and maternal tissues.

Allantoic fluid in the uterus of pregnant sows increases rapidly from day 23 of gestation until 60–65 days; amniotic fluid increases steadily from about day 30 until about day 80 and maximum growth of fetal membranes occurs between days 30 and 65. In early work with the A-mode approach, Lindahl *et al.* (1975) used a solid-state amplitude-depth ultrasonic analyser in conjunction with a 2 MHz transducer placed on the lower flank of the standing sow; echoes were displayed on an oscilloscope. An easily recognized band of echoes was obtained from a depth of 15–20 cm in pregnant sows, in contrast to echoes from a depth of only about 5 cm in non-pregnant animals. Results from about a thousand sows examined showed the accuracy of the technique to be

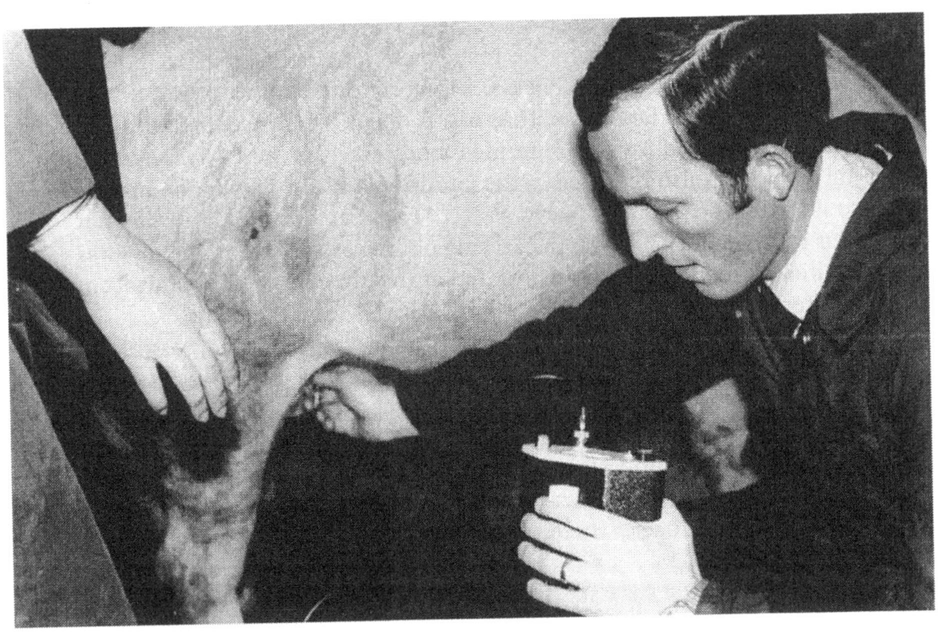

Fig. 4.2. Using A-mode ultrasonics in the detection of pregnancy in the sow.

approaching 100% for diagnosing pregnancy in sows between 30 and 90 days of gestation.

The lower limit for high accuracy in diagnosing pregnancy (30 days or so) using this technique coincides with changes in the fluid content of the uterus and the beginning of rapid growth of the fetus. The upper limit for reliability of the procedure (about 90 days) coincides with a rapid decline in the fetal fluids associated with the pig fetuses.

According to Lindahl *et al.* (1975), sows could be tested with a minimum of restraint and the equipment was capable of being operated with ease and accuracy after a short period of experience. Hansen and Christiansen (1976) employed an amplitude–depth analyser with a display system of light-emitting diodes instead of an oscilloscope; this reduced the cost and the instrument proved to be reasonably accurate. The general view in the 1970s was that A-mode rather than Doppler was the better technique to use in pigs from the viewpoint of simplicity and accuracy. There may, however, be some circumstances where it is useful to bear in mind that the upper limit of 90 days for attempting a diagnosis does not apply to the Doppler method and that A-mode sound does not provide evidence directly on the viability of fetuses.

Amplitude–depth ultrasound was evaluated between days 31 and 35 of gestation by Almond and Dial (1986); relative to the proportion of sows that farrowed, the overall accuracy of the technique was given as 86%. The same workers reported that the diode ruler amplitude–depth ultrasound had the highest rate of both false-positive and false-negative diagnoses. In Germany, Maass (1989) found the percentage of false diagnoses to be higher in gilts than in sows when using the pulse–echo device at 28 and 35 days after breeding.

4.3.3. Real-time (B-mode) ultrasound

Real-time ultrasound scanning in pregnancy diagnosis was first used on the farm in horses; subsequently, the method was applied commercially in sheep and then in cattle. In pigs, it is a matter of the cost of equipment relative to what is to be gained over the techniques already well tested. In real-time scanning, B-mode refers to brightness modality, wherein the ultrasonic image is a two-dimensional display of dots, the brightness of which is proportional to the amplitude of the returning echoes; real-time imaging refers to the moving display which is continuously presented on the monitor screen.

In France, Botero *et al.* (1984) reported on examinations carried out with scanners equipped with a 3.5 MHz transducer. The transducer was placed against the skin in the area of the abdominal flank of the standing sow or gilt, directed towards the urogenital tract and moved between the first and third posterior teats. The French workers examined 2257 sows, all presumed to be pregnant, between days 18 and 90 after insemination; the overall accuracy rate was 93.2%, with the accuracy being greater after 21 days of gestation than earlier (99.6% vs. 96.6%). Data reported by Martinat-Botte *et al.* (1985) showed that diagnosis could be made from 20 days after insemination if the

Table 4.1. The accuracy rate (%) according to the system of restraint imposed on pigs during real-time ultrasound scanning (from Martinat-Botte *et al.*, 1985).

	Day of echography after insemination					
	18–19	20–21	22–23	24–25	>26	Total
Closed building						
Tethered	90[b] (113)	93[b] (239)	96[a] (265)	95[a] (134)	96[a] (633)	95[a] (1384)
Retention box	80[c] (90)	87[c] (158)	94[a] (113)	92[c] (91)	93[c] (294)	90[c] (746)
Semi-outdoor building						
Retention box	67[c] (9)	80[c] (40)	94[a] (18)	100[a] (13)	96[a] (47)	89[c] (127)

Values are percentages with the no. of examinations in parentheses.
a vs. b, $P < 0.05$; a vs. c, $P < 0.001$.

sow was tied up during examination and 2 days later if the animal was restrained in a retention box (Table 4.1); from 22 days after mating, errors were less than 5% for positive and negative diagnoses. The French workers emphasized the need for adequate training before the operator used the technique in commercial practice.

In the UK, Jackson (1986) used a sequential linear array scanner equipped with a 3.5 MHz transducer to check 308 sows 24–37 days after service. Sows were tested in the standing position in individual pens. Vegetable oil was applied to the lower right abdominal wall as a contact agent between the animal's skin and the transducer; the transducer was placed on the right side of the sow slightly above the nipples, caudal to the navel. The author concluded that real-time scanning offered a highly accurate, rapid on-farm method of confirming pregnancy in sows from 24 days after service; routine use of the technique was seen to be unlikely, however, given the high capital cost of equipment at that time.

In Italy, Matteuzzi *et al.* (1989) reported on research to test the reliability and importance of ultrasonic scanning for early pregnancy in 3268 sows on two farms; they used two portable devices with linear-transducers either of 3.5 or 5 MHz. Optimum sensitivity and specificity were reached at 24–26 days of gestation, with 97.6% positive pregnancies as borne out by subsequent farrowings. In the same year, Thorup (1989) used real-time scanning with a 5 MHz transducer in Denmark; he found it to be a quick and relatively accurate method for pregnancy diagnosis. In the USA, Woodard *et al.* (1995) similarly evaluated the use of real-time ultrasound; using a 3.5 MHz transducer, they recorded an accuracy of greater than 90% in sows and gilts as early as day 22 of gestation.

Portable devices

A battery-operated, portable, real-time scanner for pregnancy diagnosis in pigs was tested for accuracy in 510 sows in Hungary, 21–30 days after breeding

(Szenci *et al.*, 1992). When used on days 23–30 there were eight false positives (2.5%) among 321 sows diagnosed pregnant and two false negatives (6.1%) among 33 diagnosed non-pregnant; results were less accurate on days 21–22. It was concluded that the scanner was an accurate, rapid (50–60 sows h^{-1}) method of confirming pregnancy and could be easily transported, weighing approximately 6kg including battery, video recorder and 3.5MHz sector transducer.

4.4. Oestrogen and Progesterone Assays

4.4.1. Oestrogen assays

Pigs appear to be almost unique in showing a marked increase in the concentration of oestrone sulphate in early pregnancy. Evidence was presented in the 1970s to show that the concentration of conjugated oestrogen (oestrone sulphate) increases markedly from about day 16 to a sharp peak of 2ng ml^{-1} by 25–30 days of pregnancy and then decreases to a very low level (< 0.2ng ml^{-1}) by about day 40 (Robertson and King, 1974; Hattersley *et al.*, 1980). In view of the fact that oestrone sulphate levels are negligible in non-pregnant sows, it is believed that the embryo is the source of this oestrogen; a relationship is also known to exist between the number of embryos and the concentration of oestrone sulphate in maternal plasma. For such reasons, it was suggested that measurement of this hormone at the appropriate time might be the basis of a sensitive and early pregnancy test. A competent stockperson can collect the small volumes of blood required for assay by ear vein puncture. Oestrone sulphate is known to be stable at room temperature for up to 1 week (Saba and Hattersley, 1981); shipping blood samples to a diagnostic laboratory should not present any great problem.

Several authors reported on the oestrone sulphate test in Sweden (Edqvist *et al.*, 1980), the UK (Saba and Hattersley, 1981; Cunningham *et al.*, 1983), China (Cai *et al.*, 1989) and elsewhere. According to Horne *et al.* (1983) in the USA, when a blood sample is taken at a known and specified day between days 22 and 28 and the concentration of oestrone sulphate measured, it is possible not only to distinquish between pregnant and non-pregnant sows with certainly but also to get a reasonable estimate of the number of embryos present. Stoner *et al.* (1986) recorded that concentrations of oestrone sulphate around day 30 of gestation were highly correlated with litter size at time of farrowing. In Australia, Stone *et al.* (1986) recorded significant correlations between litter size at term and oestrone sulphate in plasma on days 22 to 29 of gestation; on this basis they were able to predict litter size. In Illinois, Moenter *et al.* (1992) predicted litter size at birth with an accuracy of 73%, after measuring oestrone sulphate concentrations; litter size at birth was higher and lower than predicted in 15% and 12% of instances, respectively.

However, Atkinson *et al.* (1986) found that oestrone sulphate concentration was not an accurate predictor of litter size in their studies in Australia. Similarly, workers in the UK cautioned against attempting an accurate prediction of litter size for individual sows based on serum oestrone sulphate concentration taken during early pregnancy (Sweeney and Sutherland, 1987; Glossop and Newstead, 1987); the number of pig embryos present at days 25–30 of pregnancy may not reflect the eventual litter size because of the possibility of subsequent embryonic loss. It is also worth noting that in sows with small litter sizes, there have been occasions in which the pigs have been inaccurately assessed as non-pregnant because of low concentrations of oestrone sulphate (Almond and Dial, 1986).

In the USA, Knight and Kukoly (1990) examined the release of oestrone by the porcine placenta throughout gestation; placental oestradiol decreased sharply from 30 days to a minimum at 40 days, remained low up to 70 days and then increased to peak values at 110 days.

After about day 60 of pregnancy, further increases in the level of conjugated oestrogens occur, to be followed, from about 80–90 days of gestation, by a marked rise in free oestrogens (oestrone and oestradiol) so that a maximum of 5ng ml^{-1} plasma (total oestrogen) is reached just before term. It is believed that these free oestrogens are derived, largely if not entirely, from the fetal–placental unit.

Oestrogen in faeces

As noted above, the early pig blastocyst synthesizes substantial amounts of oestrogens (mainly oestrone and oestradiol-17β) which pass through the wall of the uterus and are sulphoconjugated to form oestrone sulphate. The presence of this metabolite in the sow's circulation, urine or faeces is a good indication of the presence of live embryos; there is a marked increase in oestrone sulphate level in the blood after day 16 of gestation with a peak being reached between days 23 and 30. The concentration of oestrogens in urine and faeces follows a similar pattern; however, faeces are more easily collected than blood or urine. In Hungary, Szenci *et al.* (1993) attempted pregnancy diagnosis on the basis of unconjugated oestrogen in faeces; faecal samples (3–5g) were taken 25–30 days after breeding. Of 146 pigs diagnosed to be pregnant after identifying oestrogen by radioimmunoassay (RIA), 137 were diagnosed correctly; an enzyme immunoassay (EIA) method was much less accurate.

4.4.2. Progesterone assays

In the 1970s, plasma progesterone levels during early pregnancy in the pig were reported by several workers. It was evident that the concentration of the steroid decreased rapidly from the mid-cycle level (25–35ng ml^{-1}) to about 10–15ng ml^{-1} at 3–4 weeks (Parvizi *et al.*, 1976); for the remainder of the gestation period, there was little further reduction in concentration. A method of early pregnancy diagnosis in the sow, based on the measurement of progesterone in the peripheral blood was described at the start of the 1970s (Robertson and Sarda, 1971). In Japan, Kawata and Fukui (1977) mentioned the difficulty of applying the method on any scale because of problems in obtaining blood samples and the expense of the laboratory procedures involved in carrying out the assay.

Some workers advocated using the progesterone test at day 18 after mating. Tests which showed plasma progesterone levels greater than 5 ng ml^{-1} at day 18 were found to be accurate in 97% of pregnant sows (Williamson *et al.*, 1980); plasma levels below this value were much less accurate in detecting non-pregnant animals because of pigs that subsequently returned to service.

According to Almond and Dial (1986), a practical limitation of the progesterone test is that blood samples must be obtained during a specific interval after mating; inaccuracies can occur when sows return to oestrus after an unsuccessful mating at a time outside the regular interval of 18–24 days after breeding. Also in circumstances in which the pregnancy is terminated after the test, as may occur after exposure to any one of several infectious agents, the sow may be assumed to be pregnant even after the litter was lost. Non-pregnant sows with multiple, luteinized cystic follicles may be anoestrous or have irregular oestrous cyclicity and be mistakenly diagnosed as pregnant based on high progesterone concentrations. Despite such limitations, Antal *et al.* (1989) in Hungary were among those who concluded that determination of serum progesterone levels was suitable for the identification of non-pregnant sows on day 22 after breeding.

Sowside progesterone tests

Use of an on-farm progesterone assay kit (Ovucheck 'Sowside') to determine pregnancy in sows was described by Glossop *et al.* (1989); blood samples were taken 17–20 days after breeding by puncturing an ear-vein with a disposable lancet. The aim was to identify empty sows at an early stage to enable them to be re-presented to the boar at the first opportunity. The authors mention problems with blood sampling on seven of 18 breeding units; although blood samples could always be obtained, the technique sometimes resulted in increased difficulty in handling animals afterwards. The apparent usefulness of the progesterone test varied among herds; it was suggested that when the rate of detection of empty sows (using the boar) was low, the progesterone test may be of greatest value.

Blood paper method of progesterone assay

Lin *et al.* (1988), in Taiwan, described their development of a blood paper method for the measurement of progesterone in blood; application of the method to early pregnancy testing of sows gave an accuracy of 90.1% for positive diagnosis and 100% for negative diagnosis in the period 17–21 days after breeding. The authors suggested that the blood paper method may provide a practical technique for use in pigs.

Progesterone in saliva
Progesterone may be measured in various body fluids, including saliva. In Japan, Moriyoshi *et al.* (1994) determined progesterone concentrations in the saliva of sows, using an enzyme immunoassay kit; pregnancy diagnosis based on progesterone concentration 17–24 days after breeding had an accuracy of 96.9% for negative results and 96.2% for those recorded as positive. The authors concluded that saliva progesterone EIA may be successfully applied in pigs as an aid to early pregnancy diagnosis.

Progesterone in faeces
In Canada, Sanders *et al.* (1994) reported studies aimed at developing a simple faecal progestin extraction and RIA method to measure immunoreactive progestin in porcine faeces; it was concluded that simple faecal progestin extraction and assay were feasible alternatives to the standard blood progesterone assays for monitoring the reproductive status of pigs. In Sweden, Hulten *et al.* (1995) reported using a faecal progesterone assay to monitor the ovulation frequency in sows in late lactation.

4.5. Other Pregnancy Testing Methods

4.5.1. Rectal palpation

The use of examination per rectum to diagnose pregnancy in the pluriparous and large primiparous sows was described by several workers during the 1960s and 1970s. The usual advice was that rectal palpation should not be attempted on a sow below about 150 kg in weight; naturally, the animal would have to be appropriately restrained. Early descriptions of the technique were generally based on the thickness and type of pulse in the middle uterine artery. Cameron (1977) showed that further information leading to an earlier and more accurate diagnosis was possible by also making a careful examination of the size, tone and position of the cervix and uterus; in most non-pregnant sows, the cervix, uterus and bifurcation of the uterus could be readily palpated and the ovaries could also be felt in many instances.

In the UK, Meredith (1976) showed that the presence of a spontaneous fremitus pulse in the middle uterine artery felt by palpation could be taken as evidence of pregnancy from 35 days after breeding to full-term. By applying digital pressure to the uterine artery, where it crossed the external iliac, fremitus was produced at an earlier stage of pregnancy; results of this work indicated an overall accuracy of 98%.

With sows in stalls or on tethers, it was found that a rectal examination took about a minute to carry out. With experience, it was evident that pro-oestrus could be detected on day 18 or 19 day after an unsuccessful mating. This led to an accurate prediction that a non-pregnant sow was due to return to oestrus. Cameron (1977) noted that when carried out at day 18 or 19 after mating or at day 30 of gestation, rectal palpation was a valuable procedure which could be used as part of a regular herd health service to intensive pig units. A comparison of rectal palpation with ultrasound (Cavalcanti *et al.*, 1989) showed that the two approaches are comparable in their accuracy. Pyorala (1989) noted that palpation per rectum is unsuitable for gilts and that other approaches are required.

4.5.2. Vaginal biopsy technique

The vaginal wall of the pig is lined by an epithelium which undergoes regular cyclical changes related to the oestrous cycle and the animal's pregnancy status. During di-oestrus, the typical vaginal epithelium is three to five cells thick; just before the heat period, rapid cell proliferation occurs and at oestrus the epithelium continues to increase in thickness and the depth may vary from five to 20 cells. This wide variation is shown by the pronounced epithelial ridging which is typical at this time. During the period just after heat, a clear reduction continues until a depth of three to five cells, typical of di-oestrus, is reached.

In the pregnant sow, examined 4–6 days after breeding, met-oestrus sloughing of excess epithelial cells occurs as in the non-pregnant animal. Until day 17 or 18, the epithelium is reduced slowly and reaches a depth of three to four cells without mitosis; at days 20–22, the vaginal epithelium of a pregnant sow has a depth of between two and four cells, occasionally five, which is in marked contrast to the ridged epithelium of oestrus. Beyond day 22 of pregnancy, the epithelium has a depth of two to three cells. Biopsy samples are taken about 5 cm caudal to the cervix with an appropriate instrument and processed according to normal histological procedures.

The diagnostic accuracy of the method from 30–90 days of pregnancy was reported in the 1960s and 1970s as greater than 90%; the obvious disadvantage of the biopsy method lies in the fact that a diagnosis cannot be made immediately because of the need for laboratory histological examination.

From a survey of the literature in the mid-1970s, Kawata and Fukui (1977) in Japan concluded that the vaginal biopsy technique was one of the most reliable and earliest methods of diagnosing pregnancy in pigs; the technique has been used in detecting sows about to return to service by taking a biopsy sample at 19 days after mating. Writing of experiences in the UK in the early 1970s, where commercial laboratories provided a service for commercial breeders, Walker (1972) stated that the method was capable of an accuracy of 95% either at about 21 days, assuming the person examining the slides was aware of the service date, or a little later in the absence of this information.

4.5.3. Prostaglandin assays

Martinat-Botte *et al.* (1980) reported on the testing of blood samples, collected 13–15 days after mating, for $PGF_{2\alpha}$. This test was based on the principle that in pregnant sows at this time $PGF_{2\alpha}$ levels are low whereas in non-pregnant pigs they are high; samples taken on day 13 resulted in an accuracy of 94% for pregnancy but only 63% for non-pregnancy. The limiting factor for development of this technique, as well as the poor prediction of non-pregnancy, was the difficulty of blood collection.

4.5.4. Immunological tests

As mentioned previously, it would be very useful to have a technique which would permit diagnosis of non-pregnancy in the sow to be made in time to allow rebreeding at 3 weeks after service. There has always been much research interest in trying to detect the presence of the very early pig embryo; Australian workers in the 1970s suggested that the detection of antigens specific to pregnancy in the pig in the early weeks might eventually result in suitable immunological tests (Cerini *et al.*, 1976). In Japan, Sekiguchi *et al.* (1992) used the rosette inhibition test, which is an assay for early pregnancy factor in blood plasma. They recorded the accuracy of diagnosis for pregnancy and non-pregnancy in the pig as 83.3% and 77.8%, respectively.

4.5.5. Diagnosis by laparoscopy

For research purposes, laparoscopy has been found to be a useful tool (Wildt *et al.*, 1975); the method can permit observation of the reproductive tract with few or no adverse physical effects either in the cyclic or pregnant pig. Very early diagnosis is possible, using luteal regression (day 12) or other features of the reproductive tract, such as hyperaemia of the uterine horns (day 15) as the guide; the method provides a means of simultaneously diagnosing pregnancy, determining ovulation rate and observing the reproductive tract.

4.6. References

Almond, G.W. and Dial, G.D. (1986) Pregnancy diagnosis in swine: a comparison of the accuracies of mechanical and endocrine tests with return to estrus. *Journal of the American Veterinary Medical Association* 189(12), 1567–1571.

Anderson, L.H., Christenson, L.K. and Ford, S.P. (1994) Estrogen associated changes in uterine blood flow and maternal recognition of pregnancy in Chinese (Meishan) pigs. *Animal Reproduction Science* 36, 135–143.

Anon. (1986) Pregnancy diagnosis. *Pig International* 16(1), 8–14.

Antal, T., Szabo, I., Soos, I., Toth, I. and Faredin, I. (1989) Comparison of the ultrasonic and serum progesterone direct RIA methods in pregnancy diagnosis of inseminated sows. *Magyar Allatorvosok Lapja* 44, 721–724.

Atkinson, S., Buddle, J.R., Williamson, P., Hawkins, C.D. and Wilson, R.H. (1986) A comparison between plasma oestrone sulphate concentration and Doppler ultrasound as methods for pregnancy diagnosis in sows. *Theriogenology* 26, 483–490.

Bazer, F.W., Ott, T.L. and Spencer, T.E. (1994) Pregnancy recognition in ruminants, pigs and horses: signals from the trophoblast. *Theriogenology* 41, 79–94.

Betteridge, K.J. and Raeside, J. (1962) Investigation of cervical mucus as an indicator of ovarian activity in pigs. *Journal of Reproduction and Fertility* 3, 410–421.

Bonnardiere, C.La, Martinat-Botte, F., Terqui, M., Lefevre, F., Zouari, K., Martai, J. and Bazer, F.W. (1991) Production of two species of interferon by Large White and Meishan pig conceptuses during the peri-attachment period. *Journal of Reproduction and Fertility* 91, 469–478.

Bosc, M.J., Martinant-Botte, F., Grouzelle, G. and Hurault, H. (1977) Bilan d'utilisation des prostaglandines au moment de la parturition chez la truie. *Journees de la Recherche Porcine en France* 9, 39–42.

Botero, O., Martinat-Botte, F. and Chevalier, F. (1984) Diagnostic precoce de gestation par echographie. *Journees de la Recherche Porcine en France* 16, 181–188.

Buddle, J.R. (1986) Pregnancy diagnosis in swine. In: Morrow, D.A. (ed.) *Current Therapy in Theriogenology.* W.B. Saunders, Philadelphia, pp. 918–923.

Cai, Z.H., Fang, Z.M., Chen, L., Guo, B.Z., Jiao, S.X. and Wang, R.X. (1989) Early pregnancy diagnosis in sows from plasma oestrone sulphate concentration. *Chinese Journal of Animal Science* 25, 12–14.

Cameron, R.D.A. (1977) Pregnancy diagnosis in the sow by rectal examination. *Australian Veterinary Journal* 53, 432–435.

Cassar, G., King, W.A. and King, G.J. (1994) Influence of sex on early growth of pig conceptuses. *Journal of Reproduction and Fertility* 101, 317–320.

Cavalcanti, S.S., Garcia, S.K. and Andreani, K.C.L. (1989) Comparative study of rectal palpation and ultrasound in pregnancy diagnosis in sows. *Arquivo Brasileiro de Medicina Veterinaria e Zootecnia* 41, 155–162.

Cerini, M., Findlay, J.K. and Lawson, R.A.S. (1976) Pregnancy-specific antigens in the sheep. *Journal of Reproduction and Fertility* 46, 65–69.

Christenson, L.K. and Ford, S.P. (1995) Comparison of PG $F_{2\alpha}$-induced luteolysis in early pregnant and estrogen-treated 'pseudopregnant' gilts. *Animal Reproduction Science* 38, 239–249.

Cross, J.C. and Roberts, R.M. (1989) Porcine conceptuses secrete an interferon during the preattachment period of early pregnancy. *Biology of Reproduction* 40, 1109–1118.

Cunningham, N.F., Hattersley, J.P. and Wrathall, A.E. (1983) Pregnancy diagnosis in sows based on serum oestrone sulphate concentration. *Veterinary Record* 113, 220–233.

Dantzer, V. (1985) Electron microscopy of the initial stages of placentation in the pig. *Anatomical Embryology* 172, 281–293.

Dhindsa, D.S., Dziuk, P.J. and Norton, H.W. (1967) Time of transuterine migration and distribution of embryos in the pig. *Anatomical Record* 159, 325–330.

Diehl, J.R. and Day, B.N. (1973) Utilization of frozen sections with the vaginal biopsy technique for early pregnancy diagnosis in swine. *Journal of Animal Science* 37, 114–117.

Diehl, J.R. and Day, B.N. (1974) Effect of prostaglandin $F_{2\alpha}$ on luteal function in swine. *Journal of Animal Science* 39, 392–396.

Done, J.T. and Heard, T.W. (1968) Early pregnancy diagnosis in the sow by vaginal biopsy. *Veterinary Record* 82, 64–66.

Dziuk, P.J. (1985) Effect of migration, distribution and spacing of pig embryos on pregnancy and fetal survival. *Journal of Reproduction and Fertility, Suppl.* 33, 57–63.

Dziuk, P.J. (1991) Reproduction in the pig. In: Cupps, P.T. (ed.) *Reproduction in*

Domestic Animals, 4th edn. Academic Press, New York, pp. 471–489.

Edgerton, L.A., Kaminski, M.A., Silvia, W.J. and Hayes, S.H. (1995) Effect of pregnancy on uterine secretion of prostaglandin $F_{2\alpha}$ in response to exogenous oxytocin in sows. *Journal of Animal Science* 73 (Suppl. 1), 212.

Edqvist, L.E., Einarsson, S. and Larsson, K. (1980) Early pregnancy diagnosis in pigs by assay of oestrone sulphate in peripheral blood plasma. *Proceedings of the International Pig Veterinary Society Congress* (Copenhagen), p. 27.

Ellicott, A.R. and Dziuk, P.J. (1973) Minimum daily dose of progesterone and plasma concentration for maintenance of pregnancy in ovariectomized gilts. *Biology of Reproduction* 9, 300–304.

Geisert, R.D., Zavy, M.T., Moffatt, R.J., Blair, R.M. and Yellin, T. (1990) Embryonic steroids and the establishment of pregnancy in pigs. *Journal of Reproduction and Fertility* Suppl. 40, 293–305.

Glossop, C.E. and Newstead, R.A. (1987) Oestrone sulphate for pregnancy diagnosis in sows. *Veterinary Record* 120, 71.

Glossop, C.E., Foulkes, J.A., Whitworth, A. and Cornwell, E. (1989) Use of an on-farm progesterone assay kit to determine pregnancy in sows. *Veterinary Record* 124, 115–117.

Grunsell, C.S. and Robertson, A. (1953) A laboratory diagnosis of pregnancy in pigs. *Veterinary Record* 65, 366–367.

Guthrie, H.D., Henricks, D.M. and Handlin, D.L. (1972) Plasma estrogen, progesterone and LH prior to estrus and during early pregnancy in pigs. *Endocrinology* 91, 675–679.

Hancock, J.L. (1961) Fertilization in the pig. *Journal of Reproduction and Fertility* 2, 307–331.

Hansen, L.H. and Christiansen, I.J. (1976) Early pregnancy diagnosis in sows by means of a newly developed ultrasonic A-scan device. *Zuchthygiene* 11, 19–21.

Hattersley, J.P., Drane, H.M., Matthews, J.G., Wrathall, A.E. and Saba, A.N. (1980) Estimation of oestrone sulphate in the serum of pregnant sows. *Journal of Reproduction and Fertility* 58, 71–72.

Horne, C., Chew, B.P., Wiseman, B.S. and Dziuk, P.J. (1983) Relationship between the level of estrone sulphate in plasma and the number of foetuses during pregnancy in the gilt. *Biology of Reproduction* 29, 56–62.

Hulten, F., Dalin, A.M., Lundeheim, N. and Einarsson, S. (1995) Ovulation frequency among sows group-housed during late lactation. *Animal Reproduction Science* 39, 223–233.

Hunter, R.H.F. (1974) Chronological and cytological details of fertilization and early embryonic development in the domestic pig, *Sus scrofa*. *Anatomical Record* 178, 169–186.

Jackson, G.H. (1986) Pregnancy diagnosis in the sow using real-time ultrasonic scanning. *Veterinary Record* 119, 90–91.

Jarrell, V.L., Day, B.N. and Prather, R.S. (1991) The transition from maternal to zygotic control of development occurs during the 4-cell stage in the domestic pig, *Sus scrofa*: quantitative and qualitative aspects of protein synthesis. *Biology of Reproduction* 44, 62–68.

Kawata, K. and Fukui, Y. (1977) Pregnancy diagnosis in the pig: a review. *Folia Veterinaria Latina* 7(2), 91–110.

King, G.J. and Ackerley, C.A. (1985) Demonstration of oestrogens in developing pig trophectoderm and yolk sac endoderm between days 10 and 16. *Journal of Reproduction and Fertility* 73, 361–367.

Klotchkov, D.V., Klotchova, A.Y., Kim, A.A. and Belyaev, D.K. (1971) The influence of photoperiodic conditions on fertility in gilts. *Proceedings of the 10th International Congress Animal Production* (Versailles), Section II, p. 8.

Knight, J.W. and Kukoly, C.A. (1990) In vitro release of progesterone and estrone by the porcine placenta throughout gestation. *Domestic Animal Endocrinology* 7, 497–508.

Knight, J.W. and Tarraf, C.G. (1994) Effect of intrauterine position on conceptus development, placental and endometrial release of progesterone and estrone in vitro throughout gestation in swine. *Biology of Reproduction* 50 (Suppl. 1), 60.

Kronnie, G.te and Boer, P.de (1993) Embryonic development in the pig up to the 64-cell stage, with reference to DNA replication and cell-cycle times from the third cleavage division. *Theriogenology* 39, 919–928.

Lin, J.H., Hwang, S.Y., Lin-Chen, Y., Wang, H.L., Wu, L.S., Hsu, T.T., Chang, S.C. and Ho, L.T. (1988) Early pregnancy diagnosis in sows by progesterone assay with blood paper method. *British Veterinary Journal* 144, 64–71.

Lindahl, I.L., Totsch, J.P., Martin, P.A. and Dziuk, P.J. (1975) Early diagnosis of pregnancy in sows by ultrasonic-depth analysis. *Journal of Animal Science* 40, 220–222.

Maass, P. (1989) Results of pregnancy diagnosis in pigs by means of ultrasound. *Tierhygiene-Information, Eberswalde-Finow* 21, 48–53.

Martin, P.A., Bevier, G.W. and Dziuk, P.J. (1977) The effect of number of corpora lutea on the length of gestation in pigs. *Biology of Reproduction* 16, 633–637.

Martinat-Botte, F., Terqui, M., Thatcher, N.W. and Mauleon, P. (1980) Early pregnancy disagnosis in the sow. *Proceedings of the International Pig Veterinary Society* (Copenhagen), p. 29.

Martinat-Botte, F., Bariteau, F., Badouard, B. and Terqui, M. (1985) Control of pig reproduction in a breeding programme. *Journal of Reproduction and Fertility* Suppl. 33, 211–228.

Matteuzzi, A., Camporesi, A. and Perazzetta, G. (1989) Ultrasonography in pregnancy diagnosis in sows: recent results. *Selezione Veterinaria* 30, 1649–1656.

Meridith, M.J. (1976) Pregnancy diagnosis in the sow by examination of the uterine arteries. *Proceedings of the International Veterinary Congress*, p. D.5.

Meridith, M.J. (1990) Pregnancy diagnosis in pigs. *Veterinary Annual* 30, 107–112.

Moenter, S.M., Webel, S.K. and Dziuk, P.J. (1992) Pregnancy detection and litter size classification by estrone sulfate measurements in swine under farm conditions. *Animal Reproduction Science* 27, 161–167.

Moryiyoshi, M., Tanaka, Y., Nakao, T. and Kawata, K. (1994) Early pregnancy diagnosis in the sow by saliva progesterone measurement using a bovine milk progesterone test EIA kit. *Journal of Reproduction and Development* 40(6), j87–j92.

Niemann, H., Illera, M.J. and Dziuk, P.J. (1983) Developmental capacity, size and number of nuclei in pig embryos cultured in vitro. *Animal Reproduction Science* 5, 311–322.

Papaioannou, V.E. and Ebert, K.M. (1988) The preimplantation pig embryo: cell number and allocation to trophectoderm and inner cell mass of the blastocyst in vivo and in vitro. *Development* 102, 793–803.

Parvizi, N., Elsaesser, F., Smidt, D. and Ellendorff, F. (1976) Plasma LH and progesterone in the adult female pig during the oestrous cycle, late pregnancy and lactation, and after ovariectomy and pentobarbitone treatment. *Journal of Endocrinology* 69, 193–203.

Perry, J.S., Heap, R.B., Burton, R.D. and Gadsby, J.E. (1976) Endocrinology of the

blastocyst and its role in the establishment of pregnancy. *Journal of Reproduction and Fertility, Suppl.* 25, 85–104.
Pope, W.F., Maurer, R.R. and Stormshak, F. (1982) Intra uterine migration of the porcine embryo: influence of estadiol-17β and histamine. *Biology of Reproduction* 27, 575–579.
Pyorala, S. (1989) Pregnancy diagnosis in swine by palpation and by amplitude-depth ultrasound scanning. *Theriogenology* 31, 1067–1073.
Roberts, R.M. and Bazer, F.W. (1988) The functions of uterine secretions. *Journal of Reproduction and Fertility* 82, 875–892.
Robertson, H.A. and King, G.J. (1974) Plasma concentrations of progesterone, oestrone, oestradiol-17β and of oestrone sulphate in the pig at implantation, during pregnancy and at parturition. *Journal of Reproduction and Fertility* 40, 133–141.
Robertson, H.A. and Sarda, I.R. (1971) A very early pregnancy test for mammals; its application to the cow, ewe and sow. *Journal of Endocrinology* 49, 407–419.
Robinson, T.J. (1957) Pregnancy. In: Hammond, J. (ed.) *Physiology of Farm Animals.* Butterworths, London, pp. 793–904.
Ryan, D.P., Yaakub, H., Harrington, D. and Lynch, P.B. (1994) Follicular development during early pregnancy and the estrous cycle of the sow. *Theriogenology* 42, 623–632.
Saba, N. and Hattersley, J.P. (1981) Direct estimation of oestrone sulphate in the sow serum for a rapid pregnancy diagnosis test. *Journal of Reproduction and Fertility* 62, 87–92.
Sanders, H., Rajamahendran, R. and Burton, B. (1994) The development of a simple fecal immunoreactive progestin assay to monitor reproductive function in swine. *Canadian Veterinary Journal* 35, 355–358.
Sekiguchi, S., Kusanagi, M., Watanabe, K. and Shimogori, Y. (1992) A study of pregnancy diagnosis making use of the rosette inhibition test in the pig. *Animal Science and Technology* 63, 310–317.
Signoret, J.P. (1980) Mating behaviour of the pig. *Reviews in Rural Science* IV, 75–78.
Simmen, R.C.M. and Simmen, F.A. (1990) Regulation of uterine and conceptus secretory activity in the pig. *Journal of Reproduction and Fertility, Suppl.* 40, 279–292.
Staigmiller, R.B. (1974) Control of luteal function during early pregnancy in the pig. *Dissertation Abstracts International* 34B, 4842–4843.
Steger, H., Kirmse, K., Loeck, G., Huhn, V., Puschel, F. and Schremmer, H. (1971) Untersuchungen zur Hell-und Dunkelstallhaltung von Zuchtsauen und Ebin. *Archiv fur Tierzucht* 14, 55–67.
Stone, B.A., Seamark, R.F., Godfrey, B.M., Quinn, P. and Lloyd, B. (1986) Oestrone sulphate levels in plasma of sows as a basis for prediction of litter size at term. *Animal Reproduction Science* 11, 51–62.
Stoner, C.S., Bazer, F.W., Thatcher, W.W., Wilcox, C.J., Combs, G.E., Knight, J.W., Wetteman, R.P. and White, C.E. (1986) Relationship between estrone sulfate in plasma and litter size at farrowing for sows and gilts. *Theriogenology* 25, 709–720.
Stork, M.G. (1979) Seasonal reproductive inefficiency in large pig breeding units in Britain. *Veterinary Record* 104, 49–52.
Sweeney, H.J. and Sutherland, H.R. (1987) Oestrone sulphate for pregnancy diagnosis in sows. *Veterinary Record* 120, 23–24.
Szenci, O., Fekete, C. and Merics, I. (1992) Early pregnancy diagnosis with a battery-operated ultrasonic scanner in sows. *Canadian Veterinary Journal* 33, 340–342.
Szenci, O., Taverne, M.A.M., Palme, R., Bertoti, B. and Merics, I. (1993) Evaluation

of ultrasonography and the determination of unconjugated oestrogen in faeces for the diagnosis of pregnancy in pigs. *Veterinary Record* 132, 510–512.

Thorup, F. (1989) Two-dimensional ultrasound scanning. Pregnancy diagnosis in sows and gilts. *Dansk Veterinaertidsskrift* 72, 433–437.

Tomes, G.J. and Nielsen, H.E. (1979) Seasonal variations in the reproductive performance of sows under different climatic conditions. *World Review Animal Production* 15, 9–19.

Torres, C.A.A. and First, N.L. (1975) Effect of surgical separation of uterus and ovaries on estrous cycle length in swine. *Journal of Animal Science* 40, 905–910.

Trout, W.E., Smith, G.W., Gentry, P.C., Galvin, J.M. and Keisler, D.H. (1995) Oxytocin secretion by the endometrium of the pig during maternal recognition of pregnancy. *Biology of Reproduction* 52 (Suppl. 1), 189.

Vallet, J.L., Klemcke, H.G. and Christenson, R.K. (1994) Effect of the conceptus and uterine space on endometrial protein secretion during mid-gestation in swine. *Journal of Animal Science* 72 (Suppl. 1)/*Journal of Dairy Science* 77 (Suppl. 1), 281.

Walker, D. (1972) Pregnancy diagnosis in pigs. *Veterinary Record* 90, 139–144.

Wiesak, T., Hunter, M.G. and Foxcroft, G.R. (1992) Ovarian follicular development during early pregnancy in the pig. *Animal Reproduction Science* 29, 17–24.

Wildt, D.E., Morcom, C.B. and Dukelow, W.R. (1975) Laparoscopic pregnancy diagnosis and uterine fluid recovery in swine. *Journal of Reproduction and Fertility* 44, 301–304.

Williamson, P. and Hennessy, D. (1975) An assessment of the vaginal biopsy technique of pregnancy diagnosis in sows. *Australian Veterinary Journal* 51, 91–93.

Williamson, P., Hennessy, D.P. and Cutler, R. (1980) The use of progesterone and oestrogen concentrations in the diagnosis of pregnancy and in the study of seasonal infertility in sows. *Australian Journal of Agricultural Research* 31, 233–238.

Woodard, T.O., Armstrong, J.D., Almond, G.W. and McCaw, M.E. (1995) Use of real-time ultasound for pregnancy diagnosis in sows and gilts. *Journal of Animal Science* 73 (Suppl. 1), 251.

5 Control of Farrowing

5.1. Introduction

In recent decades, the pig industry in many countries has tended more and more towards an intensive total confinement operation. One of the most efficient methods of operating intensive units has been by way of regular batch farrowings. In this, there is close control over all aspects of the reproductive cycle, such as weaning, breeding and birth of piglets so that farrowing batches may be maintained. However, even in well-managed units, farrowings to controlled breeding (oestrus control in cyclic gilts by hormonal agents; control in sows by weaning) can be spread out over a number of days and an effective form of inducing parturition might enable the pig producer to exercise a measure of control over this phase of operations.

Controlled parturition would enable all the piglets to be born during the normal working day or at other times, including the night, which might be considered most appropriate for ensuring the survival of as many young as possible; there is plenty of evidence to suggest that the majority of stillbirths in pigs could be prevented if a skilled stockperson could be in attendance at the time of farrowing (Fig. 5.1).

5.2. Duration of Pregnancy

It would appear that variations in breed or age of sow or in environmental effects have little influence on the length of the gestation period in the pig; the normal range can be from 110 to 120 days with commercial herd averages being around 114–116 days. The duration of pregnancy may be influenced in a minor way by genetic factors; workers in South Africa recorded a significant sire effect and others have found evidence of small but genuine differences in gestation length among breeds (Garnett and Rahnefeld, 1979). There seems to be a natural tendency for sows to farrow during the night rather than during daytime (Friend *et al.*, 1962; Bichard *et al.*, 1976).

Fig. 5.1. Heat and protection from crushing by the sow are measures used to ensure survival of the piglet.

5.2.1. Nest-building

Many authors have reported on sow behaviour in the hours leading up to farrowing; this varies according to the degree of freedom of movement enjoyed by the animal. In Australia, Haskell and Hutson (1994) showed that the distance walked by sows and gilts in the hours prior to farrowing increased significantly compared with that in previous days; nest-building occurred in the 9 h before the onset of labour. The Australian workers found no effect of parity on the level of prefarrowing movement or on the amount of nest-building performed; they took this to indicate that prefarrowing behaviour is largely unaffected by experience of farrowing and may be largely under internal control. Studies by Castren *et al.* (1993) in Finland have indicated that nest-building in sows is under multiple hormonal control and that the beginning and end of building, and the different forms of behaviour displayed by the sow, can be related to various hormonal changes. The preparturient rise in prolactin level was associated with the start of nest-building and the oxytocin concentration was related to the end of this activity.

5.2.2. Orientation of fetuses

The number of fetuses carried by the sow can have an effect on the duration of pregnancy, with small litters being carried somewhat longer than large litters. There is lack of any effect of sex ratio on gestation length in the reports of those who investigated this aspect. During prenatal life, from about day 40, it is believed that most pig fetuses face forwards and maintain a head to toe orientation with their neighbours in the uterine horn. When it comes to farrowing itself, the normal period of delivery is from 2 to 6h, with a piglet being born every 12–16 minutes, the young apparently being expelled alternately from one or other of the two uterine horns; most of the fetal membranes are passed after the last piglet is delivered (Fahmy and Friend, 1981).

5.2.3. Farrowing behaviour

A study of farrowing behaviour in Hampshire sows (Roychoudhury *et al.*, 1995) showed that about 80% of piglets were delivered head-first and the remainder were born tail-first. It is possible that a tail-first presentation may be due to a fetus passing from one uterine horn to the other and subsequently being delivered in its reversed orientation (Dziuk, 1991). A summary of the main features of the farrowing process is provided in Table 5.1.

In the monitoring of periparturient behaviour and activity in sows, researchers have used equipment of varying degrees of sophistication. In the USA, Erez and Hartsock (1990) described a microcomputer-photocell system which was used to record postural changes in crate-confined sows. The system used an infrared photocell system mounted on farrowing crates to detect changes in sow body posture before parturition; when a sow broke the infrared beam by standing or sitting, a relay connected to a microcomputer was activated.

5.3. Physiology and Endocrinology of Late Pregnancy and Parturition

Progesterone produced by the corpora lutea established at the time of mating maintains pregnancy in the pig until full-term, the steroid playing an essential role in various aspects of pregnancy, including the suppression of myometrial activity in the uterus. Much research has been devoted to elucidating the

Table 5.1. Summary of features relating to the farrowing process in the pig (from Signoret *et al.*, 1975).

Feature	Value
Vulvar swelling	1–7 days before parturition
Milky secretions on udder	6–48 h before parturition
Sow settles, lying on one side	10–90 min before parturition
Vulvar discharge	1–20 min before birth of first piglet
Time to farrow entire litter	3.5 h (0.5 to 16 h)
Additional time to expel membranes	4.5 h (0 to 12 h)
Interval time between piglets	16 min (1 min to several hours)
Piglets in posterior presentation	25–45%
Piglets in dorsosacral presentation	95%
Piglets with umbilical cord intact	60–70%
Perinatal mortality	6%
Time for piglets to suck on teats	10–35 min

factors responsible for the maintenance of the corpora lutea of pregnancy.

The onset of farrowing in the sow is preceded by a sequence of hormonal changes of both maternal and fetal origin. The events on the maternal side, which include the withdrawal of progesterone from the circulation, have been studied by many groups; changes on the fetal side, because of the greater technical difficulties involved, have been less well documented.

5.3.1. Oestrogen and progesterone levels

Dealing with maternal events, hormonal changes occur in a sequential fashion, starting with an increase in the level of oestrogens, particularly the unconjugated oestrogens, during the last 2–4 weeks of pregnancy. The increase in oestrogen concentration alone in the face of the progesterone levels found in pregnancy does not result in parturition. The progesterone concentration decreases slowly over the last 2 weeks of pregnancy but then, in the final days before farrowing, drops more quickly to reach a level of less than 2–3ng ml^{-1} plasma at the time of parturition and to less than 0.5ng ml^{-1} afterwards. Any relationship between LH and progesterone level which may operate earlier in pregnancy, no longer holds true in the last 2 days of the gestation period (Ellendorff *et al.*, 1979), indicating that this may be a period when luteotrophic support of the corpora lutea is reduced.

5.3.2. Corticosteroid levels

There is some evidence of a minor rise in the level of corticosteroids in the maternal plasma in the last 1 or 2 days of pregnancy; on the fetal side, however, the increase is proportionately greater and starts at least a week or so before parturition (Silver *et al.*, 1979).

5.3.3. Relaxin

It has been recognized for some time that the polypeptide, relaxin, is produced by the corpora lutea of the pig from day 28 to day 105 of pregnancy, is stored in the corpora lutea and mainly discharged around the time of parturition. Relaxin is known to play an important role in promoting cervical modifications essential for normal farrowing in the sow. The biophysical, biochemical and morphological effects of relaxin on the pig's cervix have been characterized. Workers have also reported on the mechanisms operating at the cellular level by which relaxin probably achieves its effect. A study by Gye-sik and Sherwood (1995) demonstrated specific relaxin binding in luminal epithelial cells, both circular and longitudinal smooth muscle cells and the blood vessels of the cervix. The authors concluded that these cell types probably contain relaxin receptors and may mediate the action of the hormone.

It is evident that the plasma concentration of relaxin rises in the days preceding parturition, with peak activity shown about 12–14h before the start of farrowing (Ellendorff *et al.*, 1979); these events are believed to coincide with changes in uterine motility from a quiescent to a labour type pattern. It is believed that relaxin and oestrogen are involved in remodelling collagen in the cervix. It is well recognized that relaxin is of considerable importance because of its effect on cervical dilation; the hormone is crucial in ensuring that the birth canal of the sow is fully dilated, thus facilitating the rapid expulsion of all the fetuses during the farrowing process.

5.3.4. Changes in components of the conceptus during pregnancy

During pregnancy in the pig, there are marked changes in the relative proportions by weight of fetal membranes, fetal fluids and fetuses (Fig. 5.2). During the early months of the gestation period, the proportion of fluids is greatest, followed by the placental membranes, then the fetuses. By the end of pregnancy, the relationship is reversed. The fetal fluids play an important role in the prenatal life of the pig, part of this being physical. By distending fetal membranes and uterine wall they contribute towards the regular spacing and protection of the developing piglets. From a little after midpregnancy, however, instead of increasing further, as in the ewe and mare, the fetal fluids in the pig are absorbed until by the end of parturition only a small volume of fluid (3% of conceptus weight) remains (Pomeroy, 1960).

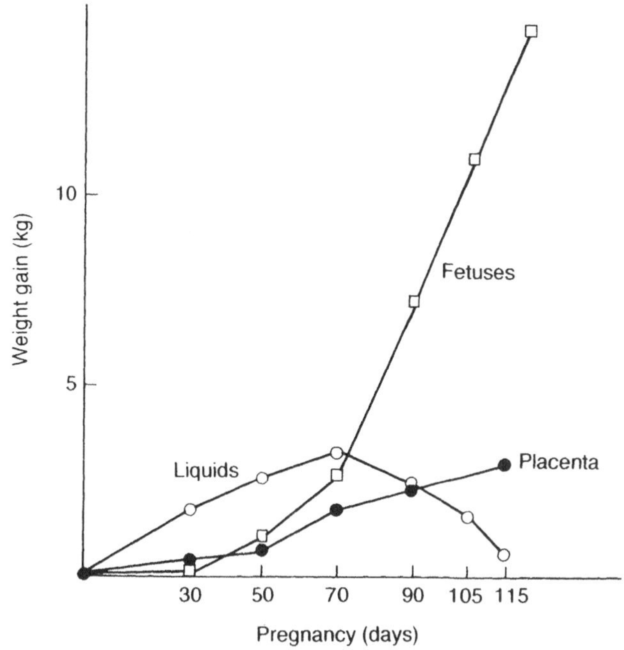

Fig. 5.2. Growth of the uterine contents of the pig during pregnancy (from Salmon-Legagneur, 1968).

5.3.5. Events at parturition

For parturition to occur, the mechanisms for the maintenance of pregnancy must cease to operate; in the case of the sow, electromyographic studies have shown that uterine motility changes rapidly from a quiescent to an active state about one-half day before the end of gestation. The changes occurring in uterine motility also coincide with nest-building behaviour in the sow and with a marked decline in the maternal progesterone level (Taverne *et al.*, 1979).

In fact, the overriding controlling factor in the onset of parturition in the sow appears to be the level of progesterone. If the concentration of the hormone declines below a certain value, farrowing can and will occur; if the progesterone level is maintained, then parturition does not occur, despite the concentrations that other hormones may reach. Experimentally, it has been shown that ovariectomy at any stage of the gestation period in pigs removes most if not all of the progesterone in circulation, resulting in abortion or 'parturition' in 24–48h (Du Mesnil du Buisson and Dauzier, 1957); on the other hand, pregnancy can be maintained in spayed pigs by administering progesterone (Ellicott and Dziuk, 1973).

It is also well established that just as the early developing pig embryos are responsible for extending the lifespan of the corpora lutea, so too, at the end of pregnancy, it is primarily a fetal effect which brings luteal function and the associated progesterone production to an end. In the pig fetus, as in the sheep, the evidence shows that the pituitary–adrenal axis has a crucial role to play in the initiation of parturition. This was demonstrated in the 1970s by the fact that fetal decapitation (Stryker and Dziuk, 1975) or fetal hypophysectomy (Bosc *et al.*, 1974) will prolong gestation beyond term; the use of corticosteroid, on the other hand, induced premature parturition (North *et al.*, 1973).

5.3.6. Fetal effects

The sequence of events at parturition in the pig is initiated by ACTH from the fetal pituitary stimulating the production of glucocorticoids by the fetal adrenals; these corticoids then apparently initiate prostaglandin production, luteal regression and the subsequent events culminating in the onset of labour. Maternal prostaglandin levels are known to remain low until farrowing begins, but then a dramatic rise occurs; in the fetal blood, the rise in prostaglandin level is earlier and coincides with the decline in progesterone (Silver *et al.*, 1979). Oxytocin concentration increases above basal levels only when plasma progesterone falls below about 10ng ml^{-1}; the concentration of this peptide undergoes a further marked increased during the expulsion phase of labour (Ellendorff *et al.*, 1979).

Wrathall *et al.* (1977) drew attention to the importance of fetal maturation on the timing of events that normally take place at the end of pregnancy; this had been demonstrated by the fact that litters in which thyroid-deficient fetuses were present tended to be farrowed late. It is well accepted that thyroid hormones play an important role in maturation processes in the brain and probably also in the pituitary–adrenal axis.

5.3.7. Fetal signals

The position of the fetus in the uterus may influence its development and the way in which the placenta secretes hormones. Studies reported by Tarraf and Knight (1995) in the USA showed that the weights of female fetuses bordered by two males and their associated membranes were lower than those from other intrauterine positions; at day 100 of gestation, placentas associated with fetuses bordered by those of the same sex released significantly more oestrone than fetuses bordered by opposite-sexed litter mates.

One area in which uncertainty existed for some time was in knowing whether there was a coordinated production of corticosteroids by all fetuses in the litter simultaneously or whether timing of parturition was triggered initially by just one or two of the more mature fetuses. Evidence showed that the presence of one or two decapitated (Stryker and Dziuk, 1975), hypophysectomized (Bosc *et al.*, 1974) or hypothyroid (Wrathall *et al.*, 1977) fetuses caused little or no delay in parturition; however, when affected fetuses outnumbered the intact ones, then pregnancy was prolonged. It was shown that in litters with a ratio of four decapitated fetuses to one intact fetus, pregnancy was prolonged; when the litter consisted of just the one intact fetus and no decapitated fetuses, the length of the gestation period was normal (Stryker and Dziuk, 1975).

As shown above, the strength of the signal to maintain the gestation period by decapitated fetuses is considerably less than that of intact fetuses trying to initiate parturition. Dziuk (1991) found that, in a litter of fetuses of varying stages of maturity, a minority of the more mature fetuses could cause the birth of the entire litter. From an evolutionary point of view, such an arrangement would favour the more rapidly developing fetuses and stabilize the length of the gestation period.

5.3.8. Phases of parturition

Although parturition in the pig is a continuous process, for descriptive purposes it is usual to divide it in to three phases: (i) the preparatory stage, in which uterine contractions commence and there is a complete dilation of the cervix; (ii) the fetal expulsion stage, which starts when the first fetus enters the birth canal (cervix and vagina); and (iii) the expulsion of fetal membranes stage.

5.3.9. Stress-induced interruption of parturition

It is known that environmental disturbance in the pig during parturition can inhibit oxytocin secretion and prolong delivery in the sow, an effect that is mediated by opioid (Lawrence *et al.*, 1992; Petherick *et al.*, 1992). The normal progress of parturition in the sow depends on the secretion of oxytocin from the posterior pituitary. The pregnant sow will, as term approaches, seek a sheltered and secluded place in which to give birth; if in the middle of parturition the sow is seriously disturbed (relocated from a familiar strawed pen to an unfamiliar farrowing crate), then oxytocin secretion is depressed and the interval between births is prolonged. It was shown, however, that administration of the opioid antagonist naloxone at such a time can restore the normal pattern of parturition, apparently by provoking a supranormal secretion of oxytocin.

In addition to the stress-induced disruption of parturition, the pain associated with parturition can also lead to increased opioid activity and may be a factor involved in the inhibition of oxytocin. Evidence of pregnancy-induced analgesia has been reported for humans and rats. Jarvis *et al.* (1996) have indicated that in the pig there is activation of an endogenous analgesic system during late pregnancy and parturition. The observed increase in pain threshold as a result of this analgesic system is believed to be mediated, at least in part, by way of an opioid mechanism during farrowing.

5.3.10. Ultrasound monitoring of parturition

Zaleski *et al.* (1995) suggested that the use of real-time ultrasound can be a powerful tool to study the progress of piglets through the reproductive tract of the sow at parturition; the arrival of piglets at the cervix, passage through the cervix and the heart-beat of the piglets could be monitored using a 5 MHz linear transrectal probe. The authors note that such a tool could play a useful part in understanding events at normal and abnormal farrowings.

5.3.11. Udder and milk effects

As described by Ash (1986), there are several signs of impending parturition in pigs. The mammary gland changes from a soft flabby structure to one of increased firmness and turgidity as parturition approaches. Oedema of the udder, presumably the result of prolactin activity, begins about 24h before delivery and milk ejection, an oxytocin-induced response, starts within 12 h prepartum.

Some workers have examined the effect of different diets fed to sows before PG-induced or progesterone-delayed farrowings. Jackson *et al.* (1995) showed that milk fat percentage was significantly increased by supplementation of the diet with 10% maize oil. The fat percentage of colostrum was reduced significantly by premature farrowing (PG on days 110 or 111) in sows on a normal diet but not in those supplemented with maize oil. In sows with delayed farrowing, the colostral immunoglobulin concentration was significantly lower on the maize oil diet than on the normal diet.

5.4. Piglet Mortality

A crucial area, in which considerable research effort has been expended, is that of improving the viability of the newborn piglet; low weight piglets and those that are weak for other reasons are the most demanding in terms of the physical environment that must be provided and the skill and effort needed by the stockperson to ensure their survival. Randall (1986) noted that, at birth, hypothermia is a major hazard to the newborn piglet, which has a limited capacity for gluconeogenesis and very little fat to act as an energy reserve. Svendsen (1992) noted how important it is to provide special care for the weak and under-sized piglets, including feeding them with a stomach tube, preferably with colostrum and milk from their dam. According to Becker (1995), recent surveys in the USA showed that the greatest percentage of neonatal deaths in pigs was due to crushing, most of this occurring in the first 3 days of life; both season of birth and sex of piglet had a significant impact on the incidence of piglet mortality.

5.4.1. Incidence of stillbirths

Evidence in the literature indicates that some 6–7% of fully formed pig fetuses are stillborn (Friend *et al.*, 1962; Randall, 1972; Dziuk, 1979); more than 80% of these stillbirths are believed to occur in the last third of the litter. Farrowings of long duration and increased time-intervals between piglets at birth are associated with a greater incidence of intrapartal stillbirths; Dziuk (1991) noted that stillborn piglets are not the small runts but often the larger fetuses in the litter. In addition to stillbirths, a further 15% or more of piglets may fail to survive the early weeks of life, with many of these dying during the first few days (Svendsen, 1992; Kuhlewind *et al.*, 1995). In Australia, a study by Stone *et al.* (1987) recorded total litter size to be a significant factor influencing the incidence of peripartum deaths.

Various categories of stillbirths

Stillbirths may be classified as prepartum deaths (Type I) or as intrapartum deaths (Type II) with intrapartum deaths making up 70–90% of all stillbirths (English and Smith, 1975). It is believed that fetal anoxia during farrowing is a major cause of Type II deaths, induced by decreased placental blood flow associated with uterine contractions, with occlusion or premature breakage of the umbilical cord, or with premature detachment of the fetal membranes. Later-born piglets are more likely to be anoxic at delivery because of the cumulative effects of prolonged contractions on the cord and fetal membranes (Randall, 1972). The incidence of Type II deaths is therefore likely to be higher in advanced parity sows in which farrowing is more prolonged (Friend *et al.*, 1962).

Weakness at birth is responsible for many losses, since it contributes to starvation, chilling, overlying and general disease. The newborn piglet, unlike the lamb, has no 'brown fat' to rely on as a heat and energy source; glycogen is the major energy reserve. Much remains to be learnt about the mechanisms controlling this glycogen supply and the way in which increases might be influenced via manipulation of the pregnant sow's diet during the final weeks of gestation.

Fraser (1990) rightly draws attention to the lack of information about the way in which the farrowing environment and the hormonal state of the sow at parturition influence the farrowing process and the vigour of the litter; factors behind grossly abnormal behaviour, such as the savaging of piglets by the sow, are poorly understood. The same author suggested that piglet mortality should be viewed as an important animal welfare issue because it is one of those areas of farming where animal welfare and commercial interests coincide (Fig. 5.3).

Large and small litters

The incidence of stillbirths is higher in litters larger than 14 and less than 5; with litters of 8, 9 and 10 piglets, very few piglets are stillborn (Dziuk, 1991). Fetuses in small litters may have to traverse regions of the uterus previously unoccupied and this may slow down their delivery; many stillbirths in large

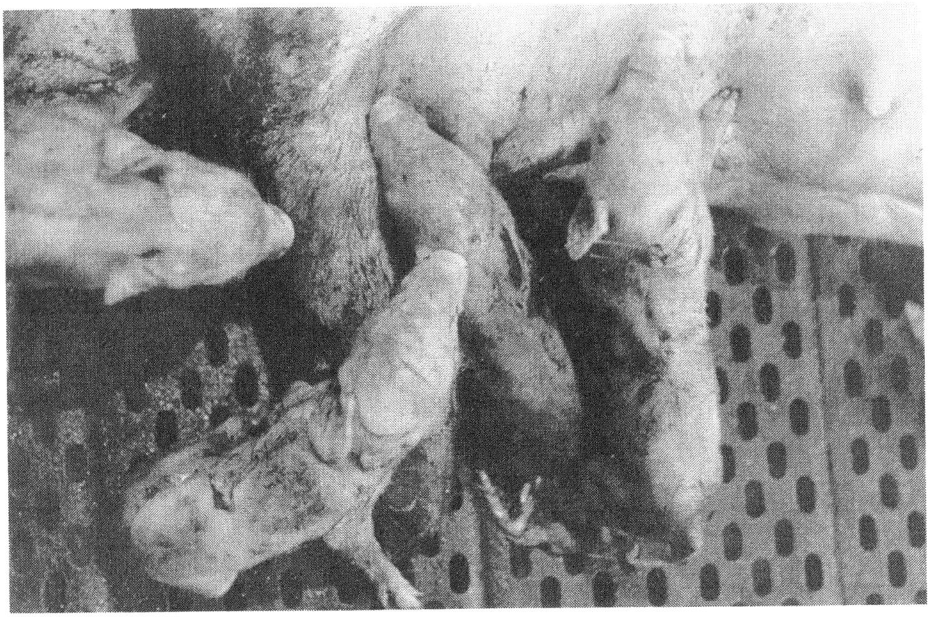

Fig. 5.3. Piglets competing for a teat in the early hours after birth following induction with prostaglandin.

litters occur when the intervals between births become extended and when the fetuses are located in the tip of the uterine horn. Fetuses can only survive in the birth canal for a few minutes after the umbilical cord is broken or compressed.

5.4.2. Relaxin and stillbirths

It has been shown, by recovering pig fetuses directly from the uterus by surgery, that nearly all the young are alive *in utero* (Dziuk, 1979), which clearly shows that the life of the piglet is seldom endangered before the birth process begins. It is, for that reason, important to understand as much as possible about parturition in the pig and the factors associated with it that may influence the extent of piglet losses. There have been indications that some stillbirths might be caused by suboptimal release of relaxin (First, 1979). Uterine contractions may not always result in the delivery of a live piglet unless the birth canal, under the influence of relaxin, is capable of dilating fully to facilitate the ready exit of the young.

5.4.3. Stockperson's role

The importance of a conscientious stockperson in minimizing piglet mortality has been emphasized in many reports. Dziuk (1979) referred to a study involving 150 supervised farrowings in which the loss of all piglets from birth to weaning was less than 6% in contrast to more than 20% in the same herd before and after the period of supervision. It has been noted that resuscitation of apparently dead piglets by an attendant is possible. Milosavljevic *et al.* (1972) mentioned a figure of 30% of apparently stillborn piglets being revived after the attendant cleared the upper respiratory tract of mucus and fluid and provided artificial respiration.

It has also been shown that survival of the newly born piglet could be aided by administering glucose by injection shortly after birth (Macpherson and Jones, 1976). Such measures can only be applied if close and immediate attention is available at the time of parturition and this certainly can be arranged more conveniently with programmed farrowings.

In the early 1980s, there was a suggestion that in many of the pig units in the UK, farrowing accommodation and the management system in operation was designed to cater for the top 85% or so of the liveborn piglets (English and Wilkinson, 1982). Management did not provide adequately for the remaining 15% of the smaller and less viable piglets. For such reasons, it is the point at which the piglet starts its decline towards death that is the important one to detect. Routine monitoring of events, both during and after farrowing, is required, to permit early correction of problems.

5.4.4. Advantages of controlled farrowings

In the normal way, farrowings occur at all hours of the night and day and gestation periods can vary in length from one sow to the next by several days. For such reasons, precise scheduling of personnel for supervision can be costly and perhaps even impossible to arrange unless the time of parturition can be closely controlled. As well as controlled farrowings to reduce piglet deaths around the time of birth, concentrated farrowings in the pig unit may offer some other potential advantages. Piglets may be more readily transferred from larger to smaller litters, and routine operations such as iron injections or teeth clipping could be carried out on larger numbers of piglets at the one time. There could be other advantages, such as reducing the age range of pigs at weaning and, by avoiding weekend and holiday farrowings, minimizing labour requirements during these high cost periods. Further details on some of the reasons for the induction of parturition in pigs are provided by Hurtgen (1986).

The benefits of induced farrowings (using cloprostenol) were outlined by Podany *et al.* (1987) after much experience of the technique in the former Czechoslovakia: (i) weekend farrowings could be avoided almost completely and the number of night farrowings could be reduced; (ii) farrowings could be attended by trained stockpersons; (iii) animal health measures, including washing of sows and disinfection of pens, can be optimized; and (iv) synchronized parturition provided conditions for more effective attendance on newborn and sucking pigs and for the production of more homogeneous groups of piglets.

In the UK, Hammond and Matty (1980) concluded that assistance of the farrowing sow by the resuscitation of weak piglets could save about one piglet per litter; for that reason alone, the control of farrowings so that the stockperson is in attendance should be an advantage. In France, Depres and Caritez (1991) discussed the behavioural and management aspects of fostering piglets on farms in which PG-synchronized farrowings had occurred. They noted that mortality rate was low among fostered piglets and that fostering increased piglet production and the number of pigs weaned per litter.

5.4.5. Oestrone levels and piglet survival

There have been indications in some studies that hormonal events during late pregannncy can be related to the subsequent viability of piglets. Work in Canada by Hacker *et al.* (1979), for example, suggested a relationship between the level of urinary oestrone secreted in late pregnancy and the viability of piglets. Under some pig farming conditions, there might be scope for using hormonal measurements to give some warning about the viability of the young which could trigger measures in the farrowing house to ensure appropriate supervision of the sow.

5.4.6. Interval between births

In the process of normal farrowing, there is evidence that the time interval between the birth of piglets is related to the incidence of stillbirths; live births occur at an average interval of 12–15 minutes, whereas the incidence of stillbirths increases as the interval from the preceding birth reaches 20 minutes (Milosavljevic *et al.*, 1972; Randall, 1972). According to Dziuk (1979), it was difficult to decide whether the stillborn piglet was the result of a long interval or whether the fetus died *in utero* and the extended interval arose from that; in view of evidence, already mentioned above, that some apparently dead pigs can be revived, Dziuk (1979) suggested that stillbirth was probably caused by the prolonged delivery rather than the other way round.

5.4.7. Prenatal treatments of the sow

There may be ways of minimizing the extent of stillbirths and perinatal deaths in pigs by specific treatments provided to the sow during pregnancy. Feeding a high energy diet (15% fat) to perinatal sows and gilts, starting on day 109 of gestation (Cieslak *et al.*, 1980) resulted in a significant increase in survival of piglets and their rate of growth in the first three weeks of life. Studies by Seerley (1989), in which sows were provided with 100 g dietary lipid supplement per kilogram from day 80 of gestation, showed that it was possible to increase markedly liver glycogen reserves in piglets at time of birth.

In the UK, Guise and Penny (1990) presented the results of a large-scale trial to determine whether iron injections 3 weeks before term had an effect on sow and piglet productivity to the point of weaning. These data indicated a small improvement in performance, estimated to be an increase in output by up to half a pig per sow per year.

It is also possible to increase muscle mass in piglets during their prenatal life. Although it is believed that muscle fibre hyperplasia ceases between day 85 and 95 of gestation in the pig (Wigmore and Strickland, 1983), it has been shown by Rehfeldt *et al.* (1993) in Germany that it is possible to increase skeletal muscle fibre number *in utero* by administration of pig somatotrophin (pST); growth hormone treatment during days 80–94 of pregnancy acclera-tated the development of the fetus and resulted in higher body weight and an advanced stage of maturity at birth.

5.4.8. Prepartal stress

In the USA, workers in the early 1980s drew attention to the fact that prepartal activity of the sow might be influenced by the type of farrowing accommodation provided; such accommodation might be a factor in explaining certain of the farrowing and lactation problems encountered in some commercial pig units in that country (Hansen and Curtis, 1980). Prior to farrowing, sows could be held in rectangular stalls with bare floors; such systems failed to take sufficient account of the sow's needs. A farrowing stall devoid of material that sows normally used to prepare a nest before parturition meant that the animal often substituted vigorous activity for nest-building a few hours before the onset of labour; such physical exertions could sometimes leave the sow with lacerations, contusions and abrasions and in a state of apparent exhaustion. It was pointed out that prepartal stress arising from the relatively barren, restraining environment of the farrowing stall could contribute towards prolonged farrowings and other problems.

In more recent times, workers in the UK have drawn attention to the fact that, in commercial practice, almost all pregnant sows are fed on restricted feeding regimes; evidence suggests that this can result in pigs being food motivated for most of the day, a behavioural state which, in a relatively barren environment, has been associated with the development of abnormal stereotypic behaviour (Spoolder *et al.*, 1995). The provision of straw may be helpful in such circumstances by buffering the adverse effects of a low feed level in group-housed pregnant sows (Spoolder *et al.*, 1996).

5.4.9. Farrowing systems and pig welfare

The welfare requirements of the sow and the litter of piglets she produces has been the subject of an increasing number of investigations in recent decades. Jensen (1996) concluded that sows should be afforded secluded nesting sites with plenty of nesting material, e.g. straw, and given the freedom to express nesting behaviour. The requirements of a commercially viable farrowing system and the housing options currently available have been discussed by Edwards and Fraser (1996), who concluded that although many alternatives to the farrowing crate have been proposed over the years, such systems have either been unsuccessful or require further development before they would be acceptable. The authors suggest that the benefits of the farrowing crate for the neonatal piglet should be acknowledged, but that systems that impose less restriction on the sow, once the piglets have established viability, should be explored.

5.5. Induction by Glucocorticoid/Oxytocin

It is well established that parturition can be induced in cattle and sheep during late pregnancy by the use of synthetic glucocorticoids. The sow, on the other hand, is not responsive to such steroid treatment until after day 100 of pregnancy and only then to relatively massive doses given on several successive days (Coggins and First, 1973). Data reported by Nara and First (1981) indicated that the mode of action of glucocorticoid treatment in termininating luteal function was not direct but was mediated through stimulation of the biosynthesis or secretion of $PGF_{2\alpha}$. The glucocorticoid approach has no place in practical pig farming.

5.5.1. Oxytocin as an induction agent

Prior to the advent of prostaglandins, there were some reports on the use of oxytocin as an induction agent. The time when this hormone proved to be effective was apparently restricted to those few hours prior to the normal start of farrowing (Costello, 1976). The general rule with oxytocin was that it only proved effective after the time that milk could be expressed from the udder. As noted later, oxytocin has since been usefully employed as a follow-up agent to prostaglandin in some induction treatments.

In passing, it may be noted that other smooth muscle stimulants, such as acetylcholine, pilocarpine, eserine and related compounds have been tried as inducers with much the same negative outcome as oxytocin (Dziuk, 1979). As mentioned elsewhere, the process of parturition involves much more than just substances which influence the contraction of the uterine myometrium.

5.6. $PGF_{2\alpha}$ and Analogues

In the early 1970s, there were many reports showing that a single intramuscular injection of $PGF_{2\alpha}$ was effective in the induction of parturition in sows treated within a few days of full-term; potent analogues of $PGF_{2\alpha}$, such as cloprostenol, were also employed with marked success. Comparisons of periparturient plasma concentrations of relaxin, progesterone and oestrogens in natural and PG-induced parturitions have failed to reveal any apparent difference in hormone patterns between the groups (King and Wathes, 1989). With natural parturition, gilts had biphasic relaxin surges, with the highest concentration occurring betweeen 12 and 28 h prepartum; PG administration produced an immediate increase in relaxin and a decline in progesterone.

5.6.1. Timing of PG administration

In many of the early induction trials, the tendency was for prostaglandin to be administered between days 108 and 113 of gestation; with experience, however, it was evident that the overall efficacy of treatment was reduced when the agent was injected more than 2 or 3 days before the usual expected date of parturiton. Under these circumstances, for breeders using PG in commercial practice, it was important to establish from their records the average gestation length for the herd in question; the usual practice was then to administer PG 2 days ahead of this herd average. It was also necessary to have accurate service dates for those sows that were to receive the agent, so that prostaglandin was only given to individual sows at the appropriate time.

5.6.2. Action of PG

The administration of $PGF_{2\alpha}$ to pregnant gilts or sows results in an immediate sharp decline in plasma progesterone levels and an associated regression of the corpora lutea. Early work also showed that PG led to the release of relaxin with peak concentrations occurring 45 minutes after the induction agent was given (Sherwood *et al.*, 1975). Numerous reports in the 1970s indicated that an orderly and normal sequence of events was initiated by PG administration. Applied under appropriate conditions, the induction treatment was seen to be without adverse effect on the health of the piglets, the duration of farrowing did not appear to be influenced and milk production remained unaffected. However, the induction of farrowings by injecting prostaglandin at day 108 could well yield piglets that were lighter in weight and less likely to survive than young produced by injection at day 111–112. Clearly, any procedure that brings about the birth of the young several days earlier than normal is likely to produce piglets of smaller than average weights; the closer to the normal full-term period that the young are born, the better their chances of survival.

5.6.3. Time interval to farrowing

It became evident that the time between PG treatment and farrowing could be influenced to a minor extent according to whether the natural prostaglandin agent or an analogue was employed. Using natural $PGF_{2\alpha}$, the interval to the onset of parturition was reported to average about 29h. Some reports showed intervals between 24 and 27h whereas others found the period to be much longer; in some instances, a proportion of the pigs which had been treated with the natural agent did not respond.

The synthetic PG analogues appeared to give a reasonably uniform response, usually showing averages of 24–28h between treatment and farrowing (Cerne and Jochle, 1981; Boland and Herlihy, 1982; Diehl and Eargle, 1985; Martin *et al.*, 1985; Itoh *et al.*, 1994). Cooper (1981) reported the peak time of farrowing as 26h after injection with the cloprostenol analogue, with 95% of the sows farrowing within 36h; in fact, 86% of sows were found to give birth in the 18-h period between 18 and 36h after treatment. The same author stressed the importance of studying herd records carefully so that the timing of induction (2 days ahead of the herd average) could be appropriately selected.

5.6.4. Comparisons between $PGF_{2\alpha}$ and analogues

Comparisons between the natural PG agent and the analogue cloprostenol were made by various workers, including some in Ireland (Table 5.2); it was usually found that the agents were similarly effective when administered two days ahead of the herd average. At Nottingham, however, Brown and Cole (1980) showed that when treatment with the natural agent and an analogue was applied on days 110, 111 and 112 of gestation, all pigs receiving the

Table 5.2. Outcome of prostaglandin treatment in the induction of parturition in the sow (from Boland *et al.*, 1979).

	I	II	III
Treatment (dose)	$PGF_{2\alpha}$ (10 mg)	Cloprostenol (175 μg)	Saline
No. of sows	27	26	28
Interval (h) to parturition	23.0 ± 10.8	24.9 ± 8.9	54.7 ± 32.8
Litter size	12.3	11.1	11.1
No. of dead at birth	0.6	1.0	2.5
Mean birthweight (lb)	2.6	2.6	2.6
Alive at 3 weeks	9.7	8.4	8.8
Mean weight at 3 weeks (lb)	10.6	10.9	10.4
Days empty	4.8	4.6	4.8

analogue responded within 48h. With the natural agent, the percentage responding within 48h increased from 29% on day 110 to 70% on day 111 to 100% on day 112; there was apparently a significant interaction between product and the day of treatment.

Transient side-effects

The transient side-effects reported after treating sows with natural $PGF_{2\alpha}$ (restlessness, increased respiration rate, salivation, tendency to urinate and defaecate, tendency towards scratching and biting) were not apparently reported to the same extent after using analogues such as cloprostenol (Widowski *et al.*, 1990). Presumably, natural PG increased smooth muscle activity to a greater extent and although such side-effects did not adversely affect the outcome of the induction treatment, the difference between agents was something to be kept in mind (Fig. 5.4). Both PG and its analogues were extensively tested before being introduced to commercial practice. The initial development work with cloprostenol, for example, involved practical evaluation in 22 major pig-producing countries. In Thailand, Chantar-aprateep *et al.* (1986a) showed that there was no difference in the response

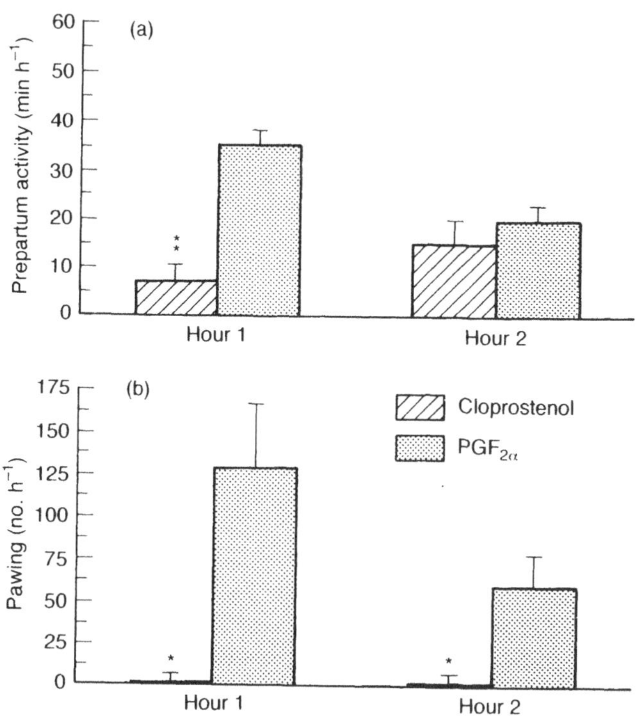

Fig. 5.4. Effect of cloprostenol (175 µg) or $PGF_{2\alpha}$ (10mg) on (a) prepartum activity (min h^{-1}) and (b) pawing (no. h^{-1}) for hours 1 and 2 after injection (from Widowski *et al.* 1990).

of sows to the intramuscular or intravenous injection of cloprostenol; clearly, the intramuscular route is the preferred approach in commercial farming.

Induction of parturition with prostaglandin, natural or analogue, was not found to affect lactation, return to oestrus after farrowing or subsequent fertility. There were indications, as discussed later, that induced farrowings might be associated with a decreased incidence of the metritis, mastitis and agalactia (MMA) syndrome in pigs. From an evaluation of more than 9000 farrowings induced by cloprostenol in the 1977–84 period in the former Czechoslovakia, Podany *et al.* (1987) concluded that the induction treatment could be used in industrial pig production without risk of affecting the reproductive performance of the sows or the quality of their offspring.

5.6.5. PG on the farm

On the basis of reports in the literature and their own experiences, King *et al.* (1979) suggested that a single injection of $PGF_{2\alpha}$, given near 08.00h in the morning should result in one-half to two-thirds of the treated pigs producing their litters during the next day; the same workers made the point that during the normal day the stockperson is engaged in routine work and might not have the time to provide constant supervision.

In an effort to improve the quality and duration of a supervision programme on an 800-sow commercial unit, Hammond and Matty (1980) induced farrowings during the night, thereby enabling the stockperson to be fully occupied attending the parturient animals and permitting daytime staff to resume supervision the following morning; effectively, this provided a period of close supervision of about 18h for each batch of farrowing sows. During the year-long trial, there was a significant increase in the average number of piglets weaned (amounting to 353 extra pigs). It was concluded that the use of controlled night farrowing offered a good opportunity under appropriate conditions of improving the supervision of farrowings and to ensure better piglet survival; an obvious limitation of the system was finding stockpersons who were willing to work night shifts.

5.6.6. Improving precision of response to PG

Use of oxytocin with PG

Several factors are now recognized as influencing the response of the sow to PG and it would be desirable in certain instances to have more precise control over the time interval between administration of the agent and the start of the farrowing process. There can certainly be a spread in the time of farrowings, which can mean that some sows may not deliver their young on the day chosen; such a delay might prevent the farmer from emptying the entire farrowing unit as planned. Oxytocin stimulates myometrial activity in the gravid uterus and had been commonly used to aid parturition in sows suffering from uterine inertia. For such reasons, there were those in the 1970s and 1980s who attempted to use oxytocin as a supplementary treatment after PG as a means of taking care of sows that otherwise would not farrow until the next day (Welk and First, 1979; Schultz, 1980; Chantaraprateep *et al.*, 1986b; Dial *et al.*, 1987; Gall and Day, 1987); this usually took the form of oxytocin injection 20–24h after the PG.

There was evidence that oxytocin given in this way led to a significant reduction in the PG to farrowing interval and permitted parturition to occur with greater synchrony. There was, however, evidence that care was necessary with the oxytocin dose level. Chantaraprateep *et al.* (1986b), for example, recorded the birth of the first piglet at an average of 21.5h after PG injection and about 1.5h after injecting oxytocin (10–20IU); use of a higher oxytocin dose (30IU) apparently caused overstimulation of myometrial activity and consequent exhaustion of the gravid uterus.

Some authors recorded the use of repeated oxytocin doses. Zarro *et al.* (1990), for example, injected sows with $PGF_{2\alpha}$ at 08.00 h on day 112 of pregnancy and 24h later gave doses of 5IU oxytocin every 30 minutes until farrowing started; they recorded about 80% of farrowings occurring during the main part of the working day (08.00 to 14.00h). The addition of oestradiol benzoate (5mg) to the induction treatment protocol has also been reported. Gall and Day (1987) recorded the most closely grouped farrowings in sows which they had treated with oestradiol after PG treatment and prior to oxytocin injection.

Not all attempts at using oxytocin after PG have been regarded as successful. In Switzerland, for example, Bilkei Papp (1994) gave sows prostaglandin analogue (3mg alfaprostol) on the morning of day 113 of pregnancy and oxytocin (10IU) 24 h later. It was concluded from the study that routine administration of oxytocin 24h after PG or after the birth of the first piglet had a harmful effect because it decreased the number of live births, possibly due to the tetanic contraction of the uterus leading to a hypoxia more severe than in the case of normal birth. It may be that a more suitable dose level of oxytocin is required.

In a general discussion of the factors affecting the time interval between PG treatment and farrowing, it might be noted that this may be shorter for large rather than small litters. In Thailand, for example, Chantaraprateep *et al.* (1986b) recorded that sows with the largest litters farrowed significantly earlier after PG treatment than other animals. These workers explained this in terms of the known evidence that parturition in the sow is largely under the influence of the pituitary–adrenal axis of the fetus; the greater the number of fetuses, the greater the effect of fetal pituitaries.

Use of xylazine (rompun) with PG

Xylazine is an alpha-2-adrenergic agonist which is known to stimulate uterine contractions in pregnant and non-pregnant pigs. Ko *et al.* (1989) in Iowa demonstrated that a cloprostenol–xylazine combination successfully induced farrowing with a high degree of synchrony; xylazine was given 20h after the PG

Table 5.3. The effect of cloprostenol–xylazine treatment on the induction of farrowing (from Ko *et al.*, 1989).

Treatment[a]	Farrowing interval (h)	Farrowing time (h)	Stillbirth (%)	Number of pigs born alive
SS	66.0 ± 8.1^{b}	5.3 ± 0.2^{b}	5.2 ± 1.7^{b}	9.4 ± 1.7^{b}
CS	10.5 ± 1.9^{c}	5.0 ± 0.2^{b}	7.3 ± 2.1^{b}	10.2 ± 2.2^{b}
SX	60.6 ± 5.6^{b}	5.2 ± 0.2^{b}	7.0 ± 2.6^{b}	9.4 ± 1.9^{b}
CX	1.5 ± 0.3^{d}	4.5 ± 0.3^{b}	6.6 ± 1.9^{b}	9.9 ± 2.3^{b}

[a]SS: saline–saline, CS: cloprostenol–saline, SX: saline–xylazine, CX: cloprostenol–xylazine ($n = 20$).
[b–d]Values with different superscript letters are significantly ($P < 0.05$) different. All values are shown as mean ± SD.

analogue (Table 5.3). The authors found that the treatment had no adverse effect on sow or piglets and allowed farrowings to be dealt with during normal working hours (07.00 to 17.00 h).

Use of carazolol with PG

The role of the autonomic nervous system in the control of parturition in pigs is not well understood. It is believed that adrenergic and cholinergic nerves are present in the myometrium and uterine vasculature of the pig. It has been suggested that the inhibition of uterine contractions during parturition, when the the sow's environment has been disturbed, may be the result of endogenous adrenaline acting through β-andrenergic receptors. A field trial was conducted by Bostedt and Rudloff (1983) to test the effect of β-adrenergic antagonists in blocking relaxation of the myometrium during farrowing and to enhance contractions. Carazolol injected after milk let-down and before the birth of the first piglet significantly reduced the duration of farrowing and the incidence of stillbirths.

In an effort to reduce variability in response to PG, Holtz and Welp (1984) tested carazolol in combination with oxytocin; carazolol decreased the interval between PG injection and parturition by about 2h and reduced the variation among sows to PG and oxytocin. In more recent times, Blanchard (1994) has described clinical trials using the drug to control prostaglandin-induced parturition; sows were given different combinations of 0–3.0mg of carazolol and 0–10IU of oxytocin. The most consistent results for length of the induction period and duration of parturition were obtained with 3.0mg carazolol alone, given 20h after the administration of prostaglandin.

Prepartum oestradiol treatment

The effect of prepartum oestradiol injection on the timing and duration of farrowing and on piglet survival was examined by Kirkwood and Thacker (1995) in Canada. Sows were given 3mg oestradiol-17β on day 112 and $PGF_{2\alpha}$ (10mg) on day 113; the time of piglet delivery was monitored between

0.800 and 18.00h on day 114 of gestation. A total of 38% of the oestrogen-treated sows farrowed before 08.00 h on day 114 compared to only 5% of control sows; oestrogen treatment had no effect on the duration of parturition (2.9h) or the interval between piglets. It was concluded that oestradiol treatment could influence the timing of onset of farrowing in some sows.

5.6.7. Shortening the farrowing process

The average duration of farrowing in sows and gilts is usually taken to be 4–6h (Friend *et al.*, 1962; Fahmy and Friend, 1981) and there is apparently no evidence to suggest that prostaglandin treatment affects the process or its duration. In view of the fact that the incidence of stillbirths increases during the last third of the farrowing period, the possibility of using agents to speed up delivery has been considered by some researchers. When uterine smooth muscle was stimulated by an appropriate agent, Leman and Sprecher (1976) obtained an extra 0.5 piglets per litter born alive by such means. This type of intervention, with the attendant time and care required for the individual sow, did not seem likely to have any commercial appeal. It seemed better to concentrate on saving the piglets when they are born.

5.6.8. Prostaglandin and the MMA syndrome

The incidence of the MMA syndrome has been reported to vary between 6 and 12%. Two main features characterize this condition; the involvement of the uterus (metritis) and the mammary gland (agalactia, mastitis). In the normal occurrence of the MMA syndrome, it has been recognized that some herds suffer from the problem more than others; this is thought to be related to the general level of health and management in the herd. Dealing with the syndrome is likely to be a matter of considering feeding, housing, health and hygiene as well as attempting specific treatment measures.

There have been indications that prostaglandin treatment for induced farrowings may help to minimize the incidence of MMA (Cerne and Jochle, 1981) by bringing about a clear-cut fall in progesterone level which might not otherwise occur as a result of the uterus failing to produce the requisite amounts of $PGF_{2\alpha}$ itself (Liptrap, 1980).

5.7. Delaying Parturition in the Pig

The synchronization of farrowing may be readily achieved by slightly shortening the normal length of the gestation period, as with prostaglandins; it may also be achieved, but much less effectively, by prolonging gestation a few days longer than usual. There may also be some circumstances in which a temporary delay in farrowing may be sought by using a tocolytic agent, such as clenbuterol, to inhibit myometrial activity.

5.7.1. Use of progesterone and progestagens

There were several reports in the 1960s and 1970s in which exogenous progesterone or synthetic progestagens were used to maintain pregnancy in the absence of normal luteal support (Sherwood *et al.*, 1978); almost invariably, fetal mortality rate was increased, often substantially, by such measures. It was apparent, however, that piglets remained alive and healthy *in utero* (shown by Caesarean section delivery of live young) for as long as 11 days beyond the expected date of parturition (Nellor *et al.*, 1975). The problem seemed to arise in the delivery of the young, presumably due to abnormal hormonal events at farrowing. In Kentucky, Wilson *et al.* (1989) studied the effect of oestradiol benzoate and $PGF_{2\alpha}$ on circulatory concentrations of progesterone and oestrogen in gilts treated with progesterone in late gestation (25 mg, four times daily from day 108 to 113); although oestrogen did not reduce progesterone levels, it significantly increased the percentage of live births. It was concluded that elevated progesterone levels were not responsible for the higher incidence of stillbirths observed with progesterone-induced delays of parturition.

When the progesterone source was luteal tissue (induced by PMSG and hCG treatment) formed a few days prior to normal term, Coggins *et al.* (1977) found that piglet mortality was similar to that in untreated controls, even when pregnancy was prolonged to 120 days or more; these workers thought that the induction of new 'accessory' corpora lutea was possibly a more physiological method of prolonging gestation in pigs than using direct progesterone treatment. Elsewhere, however, a high incidence of neonatal mortality in pigs was recorded even when parturition was delayed by way of accessory corpora lutea (Martin *et al.*, 1977).

In explaining these variable results, account was taken of evidence showing that low doses of progestagen, probably too low to inhibit the parturition process, could result in a dose-related increase in dystocia and stillbirths (Jochle *et al.*, 1974). There are also indications, even in normal farrowings, that incomplete luteal regression at parturition, with elevated maternal progesterone levels, may occur and give rise to prolonged farrowing periods and other disorders (Wrathall, 1980). Even with induced 'accessory' corpora lutea (induced CL in pigs are assumed to have a non-regressable steroidogenically active lifespan of at least 12 days), if regression did not occur in such a way that the progesterone level was sufficiently low, problems at parturition might not be unexpected. On the basis of evidence available in the 1970s, the general conclusion was that control of parturition by extending gestation beyond day 115 or shortening it to less than 112 days was unlikely to be of practical value because of unacceptable piglet mortality rates.

5.7.2. Use of synthetic progestagens

Medroxyprogesterone acetate

The synthetic progestagen medroxyprogesterone acetate (MAP), widely used in oestrus control in sheep, has been used to prolong the gestation period by up to 6 days in the sow (Curtis *et al.*, 1969). It was later evident that fetal survival was adversely affected if such MAP treatment was prolonged until day 124 in the pig (Nellor *et al.*, 1975). In Australia, treatment of sows with MAP (140mg twice daily on days 112–114) prevented farrowing even though the concentration of endogenous progesterone had declined to values equivalent to those associated with the onset of parturition in untreated animals (Whitley *et al.*, 1990). It appeared likely that MAP acted to maintain myometrial quiescence. The progestagen treatment, in this instance, was designed to lengthen the prepartum period in order to separate hormonal events normally associated with late gestation from those associated with birth and the onset of lactation in the sow.

Altrenogest

Studies with the synthetic progestagen (altrenogest) administered in oral doses of 20 mg from day 111 to day 118 were reported in the UK in the early 1980s; on withdrawal of the progestagen from the feed, sows farrowed down about 30h after the last dose (Varley, 1981). As with progesterone and MAP, lengthening gestation to 120 or 122 days by feeding altrenogest to day 118 or 120 markedly increased the incidence of stillborn piglets. Varley and Brooking (1981) recorded a stillbirth rate of 25% vs. 1.4% in controls. Studies in the USA and elsewhere, using altrenogest and PG for the control of parturition, were reviewed by Guthrie (1985).

5.7.3. Tocolytic agents

It is possible to delay farrowing by influencing the activity of the autonomic nervous system using an appropriate β_2-adrenergic agonist. Administration of the beta-agonist clenbuterol (a synthetic derivative of adrenaline; Fig. 5.5) during labour, either before birth of the first piglet or after the birth of one to three piglets, was studied by Zerobin and Kundig (1980). Injection of the agent after milk let-down but before the birth of the first piglet delayed farrowing for 15h, with no evidence of adverse effects on the eventual delivery of the young. Where labour was already in progress, it was terminated a few minutes after injection of clenbuterol and the length of delay averaged about 3h.

5.8. Induction by Progesterone Antagonist, RU 486

In the pregnant sow's uterus, progesterone binds to progesterone receptors to maintain pregnancy to full-term. In addition, progesterone is believed to

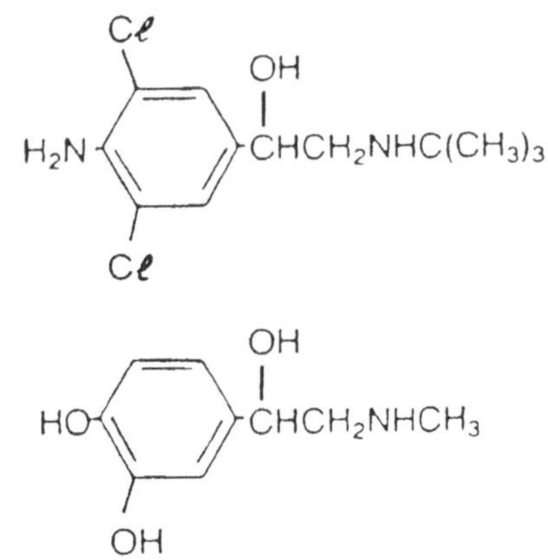

Fig. 5.5. The structural relationship of clenbuterol to adrenaline.

inhibit myometrial contractility and to ensure the quiescent state of the uterus during pregnancy. Studies in the USA have demonstrated that the progesterone antagonist RU 486, administered orally on days 111 and 112, could result in the onset of normal farrowing within 31h (Li *et al.*, 1991); the anti-progesterone brings an abrupt decline in progesterone levels and leads to the release of relaxin from the corpora lutea. The antagonist apparently renders progesterone biologically inactive and this leads to the onset of labour.

5.9. References

Ash, M. (1986) Management of the farrowing and lactating sow. In: Morrow, D.A. (ed.) *Current Therapy in Theriogenology*, W.B. Saunders, Philadelphia, pp. 931–934.

Becker, B.A. (1995) Effects of season and sex on piglet mortality. *Journal of Animal Science* 73 (Suppl. 1), 131.

Bichard, M., Stork, M.G., Rickatson, S. and Pease, A.H.R. (1976) The use of synchronized farrowing in large pig units. *Animal Production* 22, 138–139.

Bilkei Papp, G. (1994) Perinatal losses-general condition of sows. II. Effect of oxytocin given for synchronization and accelaration of farrowing. *Magyar Allastorvosok Lapja* 49, 392–395.

Blanchard, J.M. (1994) Carazolol, oxytocin, prostaglandin and induction of parturition in the sow. *Bulletin des G.T.V.* 1, 51–55.

Boland, M.P. and Herlihy, M.J. (1982) Induction of parturition in the pig using a new prostaglandin analogue (K11941). *Theriogenology* 17, 193–196.

Boland, M.P., Craig, J. and Kelleher, D.L. (1979) Induction of farrowing: comparison of the effects of prostaglandin $F_{2\alpha}$ (Lutalyse) and an analogue (cloprostenol). *Irish Veterinary Journal* 33, 45–47.

Bosc, M., du Mesnil du Buisson, F. and Locatelli, A. (1974) Mise en evidence d'uncantrole foetal de la parturition chez la truie. Interactions aved la fonction luteale. *Comptes Rendus de l'Academie Science, Paris* 278D, 1507–1510.

Bostedt, H. and Rudloff, P.R. (1983) Prophylactic administration of the beta-blocker

carazolol to influence the duration of parturition in sows. *Theriogenology* 20, 191–196.
Brown, A.C.G. and Cole, D.J.A. (1980) Prostaglandins and parturition in the sow. *Journal of Animal Science* 51 (Suppl. 1), 263.
Castren, H., Algers, B., Passille, A.M.De, Rushen, J. and Uvnas-Moberg, K. (1993) Preparturient variation in progesterone, prolactin, oxytocin and somatostatin in relation to nest building in sows. *Applied Animal Behaviour Science* 38, 91–102.
Cerne, F. and Jochle, W.F. (1981) Clinical evaluations of a new prostaglandin analogue in pigs. 1. Control of parturition and of the MMA-syndrome. *Theriogenology* 16, 459–467.
Chantaraprateep, P., Prateep, P., Lohachit, C. and Bodhipaksha, P. (1986a) Induction of parturition in sows using a prostaglandin analogue. *Australian Veterinary Journal* 63, 60–61.
Chantaraprateep, P., Prateep, P., Lohachit, C., Poomsuwan, P. and Kunavongkrit, A. (1986b) Investigation into the use of prostaglandin $F_{2\alpha}$ and oxytocin for the induction of farrowing. *Australian Veterinary Journal* 63, 254–256.
Cieslak, D.G., Leibbrandt, V.D. and Benevenga, N.J. (1980) The effect of feeding 15% fat to perinatal sows on survival and growth of piglets. *Journal of Animal Science* 51 (Suppl. 1), 59.
Coggins, E.G. and First, N.L. (1973) Response of swine and rabbits to dexamethasone at different stages of gestation. *Journal of Animal Science* 37, 305.
Coggins, E.G., Van Horn, D. and First, N.L. (1977) Influence of prostaglandin $F_{2\alpha}$, dexaethasone, progesterone and induced corpora lutea on porcine parturition. *Journal of Animal Science* 46, 754–762.
Cooper, M. (1981) Prostaglandins in veterinary practice. *In Practice* 3(1), 30–34.
Costello, W. (1976) Controlled breeding in pigs. M.Agr.Sci.Thesis, University College, Dublin.
Curtis, S.E., Rogler, J.C. and Martin, T.G. (1969) Neonatal thermostability and body composition of piglets from experimentally prolonged gestations. *Journal of Animal Science* 29, 335–339.
Depres, E. and Caritez, J.C. (1991) Fostering of pigs: behavioural and management aspects. *Productions Animales* 4, 177–182.
Dial, G.D., Almond, G.W., Hilley, H.D., Repasky, R.R. and Hagan, J. (1987) Oxytocin precipitation of prostaglandin-induced farrowing in swine: determination of the optimal dose of oxytocin and optimal interval between prostaglandin $F_{2\alpha}$ and oxytocin. *American Journal of Veterinary Research* 48, 966–970.
Diehl, J.R. and Eargle, J.C. (1985) Induced parturition in pigs with alfaprostol. *Theriogenology* 24, 655–665.
Du Mesnil du Buisson, F. and Dauzier, L. (1957) Influence de l'ovariectomie chez la truie pendant la gestation. *Comptes Rendus Seances Society Biologie Filiales* 151, 311.
Dziuk, P.J. (1979) Control and mechanics of parturition in the pig. *Animal Reproduction Science* 2, 335–342.
Dziuk, P.J. (1991) Reproduction in the pig. In: Cupps, P.T. (ed.) *Reproduction in Domestic Animals*, 4th edn. Academic Press, New York, pp. 471–489.
Edwards, S.A. and Fraser, D. (1996) Options for housing the sow and litter. *Proceedings of the British Society of Animal Science* (Winter Meeting), Paper no. 54.
Ellendorff, F., Forsling, M., Parvizi, N., Williams, H., Taverne, M. and Smidt, D. (1979) Plasma oxytocin and vasopressin concentrations in response to prostaglandin injection into the pig. *Journal of Reproduction and Fertility* 56, 573–577.

Ellicott, A.R. and Dziuk, P.J. (1973) Minimum daily dose of progesterone and plasma concentration for maintenance of pregnancy in ovariectomized gilts. *Biology of Reproduction* 9, 300–304.

English, P.R. and Smith, W.J. (1975) Some causes of death in neonatal piglets. *Veterinary Annual* 15, 95–104.

English, P.R. and Wilkinson, V. (1982) Management of the sow and litter in late pregnancy and lactation in relation to piglet survival and growth. In: Cole, D.J.A. and Foxcroft, G.R. (eds) *Control of Pig Reproduction*. Butterworths, London, pp. 479–506.

Erez, B. and Hartsock, T.G. (1990) A microcomputer-photo-cell system to monitor periparturient activity of sows and transfer data to a remote location. *Journal of Animal Science* 68, 88–94.

Fahmy, M.H. and Friend, D.W. (1981) Factors influencing the repeatability of the duration of farrowing in Yorkshire sows. *Canadian Journal of Animal Science* 61, 17–22.

First, N.L. (1979) Mechanisms controlling parturition in farm animals. *Proceedings of the Animal Reproduction Symposium*. Allenheld, Osmun, Montclair, pp. 215–257.

Fraser, D. (1990) Behavioural perspectives on piglet survival. *Journal of Reproduction and Fertility* Suppl. 40, 355–370.

Friend, D.W., Cunningham, H.M. and Nicholson, J.W.G. (1962) The duration of farrowing in relation to reproduction performance of Yorkshire sows. *Canadian Journal of Comparative Medicine* 26, 127–182.

Gall, M.A. and Day, B.N. (1987) Induction of parturition in swine with prostaglandin $F_{2\alpha}$, estradiol benzoate and oxytocin. *Theriogenology* 27, 493–505.

Garnett, I. and Rahnefeld, G.W. (1979) Factors affecting gestation length in the pig. *Canadian Journal of Animal Science* 59, 83–87.

Guise, H.J. and Penny, R.H.C. (1990) Influence of supplementary iron in late pregnancy on the performance of sows and litters. *Veterinary Record* 127, 403–405.

Guthrie, H.D. (1985) Control of time of parturition in pigs. *Journal of Reproduction and Fertility* Suppl. 33, 229–244.

Gye-sik Min and Sherwood, O.D. (1995) Identification of specific relaxin-binding cells in the cervix of the pregnant pig utilizing a biotinylated porcine relaxin probe. *Biology of Reproduction* 52 (Suppl. 1), 101.

Hacker, R.R., Hazeleger, W., Van Poppel, F.J.J., Osinga, A., Verstegen, M.W.A. and Van de Wiel, D.F.M. (1979) Urinary oestrone concentration in relation to piglet viability, growth and mortality. *Livestock Production Science* 6, 313–318.

Hammond, D. and Matty, G. (1980) A farrowing management system using cloprostenol to control the time of parturition. *Veterinary Record* 106, 72–75.

Hansen, K.E. and Curtis, S.E. (1980) Prepartal activity of sows in stall or pen. *Journal of Animal Science* 51, 456–460.

Haskell, M.J. and Hutson, G.D. (1994) Pre-farrowing behaviour of sow and gilts with access to space for locomotion. *Australian Journal of Experimental Agriculture* 34, 1099–1105.

Holtz, W. and Welp, C. (1984) Induction of parturition in sows by prostaglandin–oxytocin programs. *Proceedings of the 8th International Pig Veterinary Society Congress* (Ghent), p. 378.

Hurtgen, J.P. (1986) Induction of parturition in swine. In: Morrow, D.A. (ed.) *Current Therapy in Theriogenology*. W.B. Saunders, Philadelphia, p. 927.

Itoh, S., Sone, M., Suzuki, K., Izumiya, K., Mochizuki, H., Tsutsui, T., Nakazima, C., Ogasa, A. and Nakahara, T. (1994) Induction of parturition in pigs with a

prostaglandin $F_{2\alpha}$ analogue (Fenprostalene). *Animal Science and Technology* 65, 834–841.
Jackson, J.R., Hurley, W.L., Easter, R.A., Jensen, A.H. and Odle, J. (1995) Effects of induced or delayed parturition and supplemental dietary fat on colostrum and milk composition in sows. *Journal of Animal Science* 73, 1906–1913.
Jarvis, S., McClean, K.A., Calvert, S., Chirnside, J., Deans, L., Lawrence, A.B. and Molony, V. (1996) Changes in pain threshold during pregnancy and parturition in the sow and the involvement of opioids. *Proceedings of the British Society of Animal Science* (Winter Meeting), Paper no. 32.
Jensen, P. (1996) The welfare requirements of the sow and litter. *Proceedings of the British Society of Animal Science* (Winter Meeting), Paper no. 53.
Jochle, W., Orozoo, L., Zerobin, K., Esparaza, H. and Hidalge, M.A. (1974) Effects of a progestin and parturition and the post-partum period in pigs. *Theriogenology* 2, 11–20.
King, G.J., Robertson, H.A. and Elliott, J.I. (1979) Induced parturition in swine herds. *Canadian Veterinary Journal* 20, 157–160.
King, G.J. and Wathes, D.C. (1989) Relaxin, progesterone and estrogen profiles in sow plasma during natural and induced parturitions. *Animal Reproduction Science* 20, 213–220.
Kirkwood, R.N. and Thacker, P.A. (1995) Effects of pre-partum oestradiol injection on parturition in sows and piglet survival. *Animal Science* 60, 481–483.
Ko, J.C.H., Evans, L.E., Hsu, W.H. and Hopkins, S.M. (1989) Farrowing induction with cloprostenol–xylazine combination. *Theriogenology* 31, 795–800.
Kuhlewind, J., Mewes, I., Horugel, K. and Kohler, U. (1995) Management of piglet production. *Neue Landwirtschaft* no. 10, 61–66.
Lawrence, A.B., Petherick, J.C., McClean, K.A., Gilbert, C.L., Chapman, C. and Russell, J.A. (1992) Naloxone prevents interruption of parturition and increases plasma oxytocin following environmental disturbance in parturient sows. *Physiology and Behaviour* 52, 917–923.
Leman, A.D. and Sprecher, D.J. (1976) Effect of parasympathomimetics on farrowing and stillbirth rate. *Proceedings of the International Pig Veterinary Society* (Ames), p. D.20.
Li, Y., Huang, C.J., Klindt, J. and Anderson, L.L. (1991) Divergent effects of antiprogesterone, RU 486, on progesterone, relaxin and prolactin secretion in pregnant and hysterectomized pigs. *Endocrinology* 129, 2907.
Liptrap, R.M. (1980) Prostaglandin $F_{2\alpha}$ and progesterone in experimental hypogalactia in sows. *Research in Veterinary Science* 29, 240–247.
Macpherson, R.M. and Jones, A.S. (1976) The effect of administration of glucose on survival of the neonatal pig. *Proceedings of the 62nd Meeting of the British Society of Animal Production* (Winter Meeting), Paper no. 38.
Martin, P.A., Norton, H.W. and Dziuk, P.J. (1977) The effects of corpora lutea induced during pregnancy on the length of gestation in the pig. *Biology of Reproduction* 17, 712–717.
Martin, M.J., Meisinger, T.C., Flowers, W.L., Cantley, T.C. and Day, B.N. (1985) Parturition control in sows with a prostaglandin analogue (alfaprostol). *Theriogenology* 24, 13–19.
Milosavljevic, S., Miljkovic, V., Soljanski, B., Radovid, B., Trbojevic, C. and Stankov, M. (1972) The revival of apparently stillborn piglets. *Acta Veterinaria (Beograd)* 22, 71–76.
Nara, B.S. and First, N.L. (1981) Effect of indomethacin on dexamethasone-induced

parturition in swine. *Journal of Animal Science* 52, 788–793.
Nellor, J.E., Daniels, R.W., Hoefer, J.A., Wildt, D.E. and Dukelow, W.R. (1975) Influence of induced delayed parturition on fetal survival in pigs. *Theriogenology* 4, 23–31.
North, S.A., Hauser, E.R. and First, N.L. (1973) Induction of parturition in swine and rabbits with the corticosteroid dexamethasone. *Journal of Animal Science* 36, 1170–1174.
Petherick, J.C., Lawrence, A.B. and McLean, K. (1992) Stress-induced disruption of parturition in gilts. *Journal of Animal Science* 70 (Suppl. 1), 162.
Podany, J., Majerciak, P., Stejskal, J. and Stefan, M. (1987) Fertility of sows with cloprostenol-induced parturition in an all-in-all-out management system. *Pig News and Information* 8, 23–27.
Pomeroy, R.W. (1960) Infertility and neonatal mortality in the sow. *Journal of Agricutural Science (Cambridge)*, 54, 1–56.
Randall, G.C.B. (1972) Factors influencing stillbirth and perinatal mortality. *Veterinary Record* 90, 183–186.
Randall, G.C.B. (1986) Physiology of late pregnancy and parturition in swine. In: Morrow, D.A. (ed.) *Current Therapy in Theriogenology.* W.B. Saunders, Philadelphia, pp. 923–926.
Rehfeldt, C., Fiedler, I., Weikard, R., Kanitz, E. and Ender, K. (1993) It is possible to increase skeletal muscle fibre number in utero. *Bioscience Reports* 13, 213–220.
Roychoudhury, R., Sarker, A.B. and Bora, N.N. (1995) Farrowing and suckling behaviour of Hampshire pigs. *Indian Journal of Animal Production and Management* 11, 62–64.
Salmon-Legagneur, E. (1968) Prenatal development in the pig and some other multiparous animals. In: Lodge, G.A. and Lamming, G.E. (eds) *Growth and Development of Mammals.* Butterworths, London, pp. 158–191.
Schultz, R.H. (1980) Experiences and problems associated with usage of prostaglandins in countries other than the United States. *Journal of the American Veterinary Medical Association* 176, 1182–1186.
Seerley, R.W. (1989) Survival and post weaning performance of pigs from sows fed fat during late gestation and lactation. *Journal of Animal Science* 67, 1889–1894.
Sherwood, O.D., Chang, C.C., De Vier, G.W., Diehl, J.R. and Dziuk, P.J. (1975) Relaxin concentrations in pig plasma following the administration of prostaglandin $F_{2\alpha}$ during late pregnancy. *Endocrinology* 98, 875–879.
Sherwood, O.D., Wilson, M.E., Edgerton, L.A. and Chang, C.C. (1978) Serum relaxin concentrations in pigs with parturition delayed by progesterone administration. *Endocrinology* 102, 471–476.
Signoret, J.P., Baldwin, B.A., Fraser, D. and Hafez, E.S.E. (1975) The behaviour of swine. In: Hafez, E.S.E. (ed.) *The Behaviour of Domestic Animals*, 3rd edn. Baillière Tindall, London, pp. 295–329.
Silver, M., Barnes, R.J., Comline, R.S., Dowden, A.L., Clover, L. and Mitchell, M.D. (1979) Prostaglandins in the fetal pig and prepartum endocrine changes in mother and fetus. *Animal Reproduction Science* 2, 305–322.
Spoolder, H.A.M., Burbidge, J.A., Lawrence, A.B., Simmins, P.H. and Edwards, S.A. (1995) Provision of a foraging substrate reduces the development of excesssive chain and bar manipulation in food restricted sows. *Applied Animal Behaviour Science* 43, 249–262.
Spoolder, H.A.M., Burbidge, J.A., Edwards, S.A., Lawrence, A.B. and Simmins, P.H. (1996) Effects of food level and straw provision during pregnancy on sow

performance and responses to an adrenocorticotrophic hormone challenge. *Proceedings of the British Society of Animal Science* (Winter Meeting), Paper 197.

Stone, B.A., Heap, P.A. and Godfrey, B.M. (1987) Induction of farrowing in sows with prostaglandin $F_{2\alpha}$, alone or in combination with a parasympathomimetic or oxytocin. *Australian Veterinary Journal* 64, 254–256.

Stryker, J.L. and Dziuk, P.J. (1975) Effects of fetal decapitation on fetal development, parturition and lactation in pigs. *Journal of Animal Science* 40, 282–287.

Svendsen, J. (1992) Perinatal mortality in pigs. *Animal Reproduction Science* 28, 59–67.

Tarraf, C.G. and Knight, J.W. (1995) Effect of intrauterine position on conceptus development, placental and endometrial release of progesterone and estrone in vitro, and concentration of steroid hormones in fetal fluids throughout gestation in swine. *Domestic Animal Endocrinology* 12, 179–187.

Taverne, M., Willemse, A.H., Dieleman, S.J. and Bevers, M. (1979) Plasma prolactin, progesterone and oestradiol-17β concentrations around parturition in the pig. *Animal Reproduction Science* 1, 257–283.

Varley, M. (1981) New ways to control farrowing. *Pig Farming* (August issue), pp. 38–40.

Varley, M.A. and Brooking, P. (1981) The control of parturition in the sow using an oral progestagen. *Animal Production* 32, 369.

Welk, F. and First, N.L. (1979) Effect of oxytocin on the synchrony of parturition induced by $PGF_{2\alpha}$ (ICI 80996). *Journal of Animal Science* 49 (Suppl. 1), 347.

Whitely, J.L., Hartmann, P.E., Willcox, D.L., Bryant-Greenwood, G.D. and Greenwood, F.C. (1990) Initiation of parturition and lactation in the sow: effects of delaying parturition with medroxyprogesterone acetate. *Journal of Endocrinology* 124, 475–484.

Widowski, T.M., Curtis, S.E., Dziuk, P.J., Wagner, W.C. and Sherwood, O.D. (1990) Behavioural and endocrine responses of sows to prostaglandin $F_{2\alpha}$ and cloprostenol. *Biology of Reproduction* 43, 290–297.

Wigmore, P.M.C. and Stickland, N.C. (1983) DNA, RNA and protein in skeletal muscle of large and small pig fetuses. *Growth* 47, 67–76.

Wilson, M.E., Edgerton, L.A., Stahly, T.S. and Cromwell, G.L. (1989) Effects of estrogen and prostaglandin on progesterone-delayed farrowing. *Theriogenology* 32, 577–583.

Wrathall, A.E. (1980) Ovarian disorders in the sow. *Veterinary Bulletin* 50, 253–272.

Wrathall, A.E., Bailey, J., Wells, D.D. and Herbert, C.N. (1977) Studies on the barker (neonatal respiratory distress) syndrome in the pig. *Cornell Veterinarian* 67, 543–598.

Zaleski, H.M., Dewey, C.E. and Dobson, H. (1995) A method to identify problems in the passage of piglets through the reproductive tract at parturition. *Biology of Reproduction* 52 (Suppl. 1), 190.

Zarro, E., Mandarino, P. and Kennett, D.L. (1990) Results of an experiment on programmed parturition in pigs. Use of oxytocin in repeated low doses. *Revista di Suinicoltura* 31, 97–100.

Zerobin, K. and Kundig, H. (1980) The control of myometrial functions during parturition with a beta 2-mimic compound, planipart. *Theriogenology* 14, 21–35.

6 More Frequent Farrowings in Pigs

6.1. Introduction

Sow productivity is closely related to the number of weaned pigs produced per sow per year; this, in turn, is a function of the number of piglets born and reared per litter and the number of litters per sow per year. Management and husbandry techniques can influence the number of piglets born. The level of feeding can be increased between weaning and remating, particularly for first and second litter sows; older animals will also benefit in the same way if they have lost weight excessively during lactation. Results are generally better with crossbred animals, which are known to produce larger litters and more viable piglets.

Beyond factors such as those, however, there is one major way in which sow productivity can be influenced; this is by reducing the age of weaning. Decreasing the length of the lactation period by early weaning can increase the number of litters per sow per year from 2.05, when piglets are weaned at 8 weeks to 3.0 when the sow is weaned at birth, assuming an interval of a week between weaning and conception. The effect of age of weaning on the possible number of pigs reared annually has been discussed by various authors. In the UK, Newport (1977) calculated that, compared with weaning at 21 or 56 days of age, weaning at birth could increase sow productivity by about 40% and 70%, respectively.

Although in theory it may appear that the sow is capable of producing three litters per year, there are, in practice, several difficulties facing the producer who attempts to push the sow towards that limit of performance. The sow, in contrast to the cow, is normally in a state of anoestrus during lactation and for that reason cannot normally be bred to the boar until some days after the piglets are weaned. It follows, therefore, that methods of overcoming lactational anoestrus could be of some importance in developing techniques for increasing the number of litters produced annually by the sow.

6.2. Physiology and Endocrinology of the Lactational Anoestrus in Pigs

6.2.1. Early postpartum heats

With the exception of the anovulatory heat period that may occur in a proportion of sows in the early days after farrowing (Baker *et al.*, 1953), oestrous activity is normally inhibited during lactation in the sow. It is believed that the anovulatory oestrus shown after farrowing in pigs may be a result of the high levels of placental oestrogens present in the sow prior to parturition. It is evident that lactational anoestrus is maintained until weaning, provided that more than one sucking piglet is continuously present (Parvizi *et al.*, 1976) and that the lactation itself is not unduly prolonged.

6.2.2. Inhibition of the hypothalamic–pituitary–ovarian axis

The underlying endocrine mechanisms at work during lactation in the sow are not well understood. It is clear that ovarian and pituitary activity is suppressed while the sow continues to nurse the piglets; follicular growth and corpora lutea are absent during the entire lactation period, plasma progesterone levels are consistently low and plasma oestrogen concentrations may be almost undetectable.

LH levels

Pituitary LH content and plasma LH levels are very low during lactational anoestrus; in some studies, there were indications that the anterior pituitary may be unable to release LH during lactation because the stimulatory positive oestrogen feedback mechanism may be incapable of operating normally (Elsaesser and Parvizi, 1980).

Results of work by Renesis *et al.* (1993a) have suggested that the initial development of the well-characterized inhibitory effect of suckling on LH secretion in sows may not be opioid-dependent but that dependency on an opiate-dependent mechanism may become an important component of LH suppression in established lactation. In a further paper, Renesis *et al.* (1993b) concluded that the development of lactational anoestrus in sows requires suckling-induced inhibition of LH secretion by 78h postpartum; this inhibition of LH release did not appear to be causally related to short-term changes in prolactin secretion.

In a review of the endocrine bases of lactational anoestrus in the sow, Quesnel and Prunier (1995) note that follicular development and LH secretion are depressed during the last month of pregnancy and that, after farrowing, LH secretion increases and is then inhibited again by the establishment of lactation. Lactating sows then have to contend with the suckling stimuli originating from the piglets and the high nutrient requirements for milk production. As lactation proceeds, LH secretion increases and there is a further rise in gonadotrophin concentration at weaning.

FSH levels

Variations in FSH levels are less evident than those of LH, and it is believed that this is because FSH secretion depends mainly on ovarian inhibition whereas LH secretion depends on factors relating to lactation. The development of follicles progressively resumes during lactation and these follicles acquire the ability to respond to hormonal events at weaning-time to begin preovulatory growth. Hormones involved in lactation, such as prolactin, insulin, growth hormone and insulin-like growth factor I, may influence follicle development at the ovarian level or by influencing the secretion of gonadotrophins. Quesnel and Prunier (1995) concluded that inhibition of the hypothalamic–pituitary–ovarian axis during lactation in the pig is mainly due to suckling-induced neuroendocrine reflexes and that nutritional deficit becomes relatively more important during the third and fourth weeks after farrowing (Fig. 6.1).

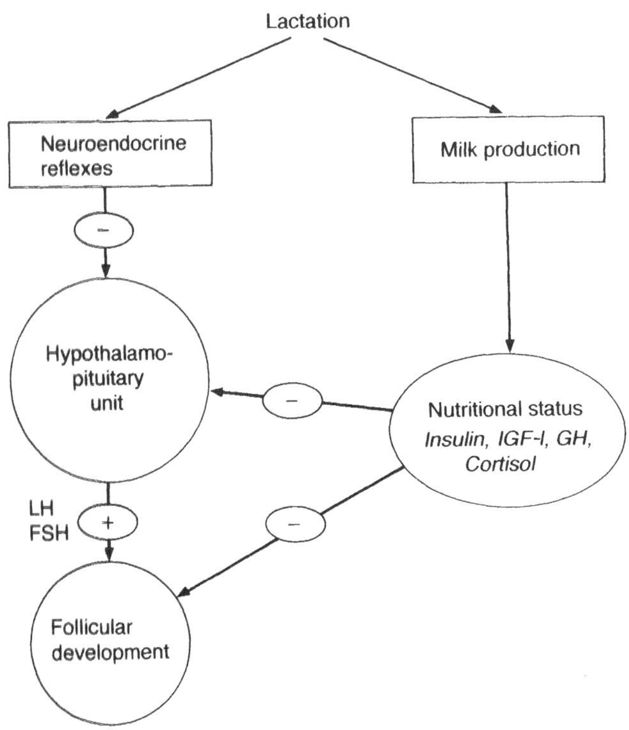

Fig. 6.1. Factors influencing the secretion of gonadotrophins in the lactating sow (from Quesnel and Prunier, 1995).

6.2.3. First parity sows

The pig is a good model for demonstrating nutrition–reproduction interactions (Foxcroft and Cosgrove, 1994). Tissue catabolism during lactation invariably reinforces the suckling-induced suppression of LH secretion in first parity sows; relationships are well established between insulin status, LH secretion and postweaning fertility in this species. Foxcroft and Cosgrove (1994) note that limited fat reserves during lactation and the partitioning of nutrients for lean tissue growth after weaning, are likely to be major factors in reduced fertility in primiparous pigs; nutritionally mediated effects on the extent of early embryo mortality in pigs have been shown to be associated with differences in progesterone levels in the immediate postovulatory period.

In Australia, Paterson and Pearce (1994) examined the hypothesis that the interval from weaning to oestrus in primiparous sows was controlled by LH secretion at or around the time of weaning; they were able to show that changes in circulating LH associated with weaning were critical to the resumption of cyclic activity. In investigating some of the causes of reduced fertility in first parity sows, the effect of insulin administered to primiparous sows during postweaning follicular growth on subsequent litter size was examined by Ramirez *et al.* (1994). They concluded that insulin treatment prior to breeding alleviated the depressed litter size characteristic of second litter sows without adversely affecting later reproductive performance.

At a more practical level, Sterning *et al.* (1994) examined the duration and intensity of external oestrous symptoms in primiparous sows at the first and second heats after weaning. The Swedish workers reported a positive correlation between the duration of pro-oestrus and the intensity of the reddening and swelling of the vulva at the first oestrus after weaning.

6.2.4. Prolactin levels

It is likely that the release of prolactin, as a result of the suckling stimulus, plays an important role in suppressing the stimulatory oestrogen feedback action on gonadotrophin secretion. Dusza and Krzymowska (1981) reported on prolactin levels in sows during pregnancy, parturition and the early weeks of lactation. They recorded increases in prolactin just before farrowing and a very high level at the time of parturition (147ng ml^{-1}); after farrowing, the level gradually decreased, reaching 43ng ml^{-1} on day 5 of lactation. At weaning, prolactin levels decline markedly (Edwards and Foxcroft, 1983). Li *et al.* (1991) have also shown that prolactin concentrations in peripheral plasma are high at parturition, increase in response to sucking by the piglets and decline after weaning. The antigonadal role of prolactin is well established in other species and lactating pigs are known to have high prolactin levels in comparison with sows in the postweaning period. An impairment of LH release after the administration of oestrogen to hyperprolactinaemic sheep was reported, and it appears that this was not due to a reduced responsiveness of the pituitary to GnRH (Kann *et al.*, 1976).

If prolactin also plays an important role in lactational anoestrus in pigs, the recovery of oestrogen-induced LH release during late lactation might be explicable on the basis of the progressive decrease of plasma prolactin levels in the period between farrowing and weaning (Landeghem and Wiel, 1978). The gradual decline in prolactin secretion in late lactation is apparently due to a decrease in suckling frequency, rather than any decline in prolactin level at the time of suckling (Varley and Foxcroft, 1990). In the rat, it is evident that the suckling stimulus rather than high circulating levels of prolactin is primarily responsible for suppressing LH release; in the sow, as a result of increased solid intake by piglets, it is possible that there may be a partial recovery of the stimulatory oestrogen feedback mechanism.

6.2.5. Involution of the uterus

In the matter of rebreeding the sow in the early weeks after farrowing, one obvious consideration, quite apart from the question of ovarian activity, is that of involution of the uterus. According to the literature, in conventional weaning systems, the sow's uterus rapidly loses length and weight in the first 2–3 weeks after farrowing and then remains in that state until after weaning, when its weight and length increase in response to ovarian hormones; although complete involution does not occur until about 3 weeks into lactation, it seems that most of the important changes have taken place by day 7 of lactation and it appears that the uterus should be capable of functioning normally again by 2–3 weeks after parturition (Palmer *et al.*, 1965; Svajgr *et al.*, 1974).

It should be noted, however, that the rate of uterine involution is probably influenced by whether or not the sow suckles after farrowing; there have been reports that very short lactations may slow down the rate of loss in weight of the postpartum uterus. If mating occurs very soon after parturition, then conditions in the reproductive tract may not be particularly favourable for fertilization and the early development of embryos.

6.3. Initiating Ovarian Activity in the Postpartum Sow

Based on experimental evidence, the sow ovary is apparently responsive to stimulation by exogenous gonadotrophins during lactation, but does not seem capable of secreting significant amounts of steroids until after weaning (Stevenson and Britt, 1981); ovulation during lactation is apparently inhibited due to reduced LH synthesis or secretion that can only be increased by weaning or perhaps a reduction in suckling intensity. Removal of the litter and weaning of the sow leads to a transient increase in plasma LH level, a gradual increase in plasma oestradiol concentrations and finally in the preovulatory surge of LH that occurs at the first postweaning oestrus.

6.3.1. Management practices

The normal postweaning oestrus

In many mammalian species, parturition is followed by a period of anovulation of variable duration. Lactation is one of the main regulators, with a strong inhibitory influence on the occurrence of oestrus and ovulation. In wild boar herds or where domestic sows are living under semi-natural conditions, weaning of piglets is a slow and gradual process extending over many weeks; in commercial pig production, however, lactation is interrupted at 3–5 weeks after farrowing, which is about the time of peak milk production.

The effect of lactation on suppressing ovarian activity in the sow stems primarily from suckling stimuli but there is also the metabolic burden of milk production itself to consider (Fig. 6.2). It is well established that removal of the piglets from the sow after 3 to 5 weeks of lactation should result in oestrus and ovulation within 4 to 8 days (Maurer *et al.*, 1985). However, as noted in may reports, a significant source of loss in commercial pig production stems from the failure of some portion of the sows to show oestrus at the expected time.

6.3.2. Seasonal factors

In the USA, Britt *et al.* (1983) noted that although sows of all parities were affected, postweaning anoestrus was more prevalent in primiparous sows during the summer and autumn seasons. It is believed that the difficulty lies in insufficient stimulation of the ovaries, which is probably a reflection of inadequacies in the workings of the hypothalamic–pituitary axis. Studies by

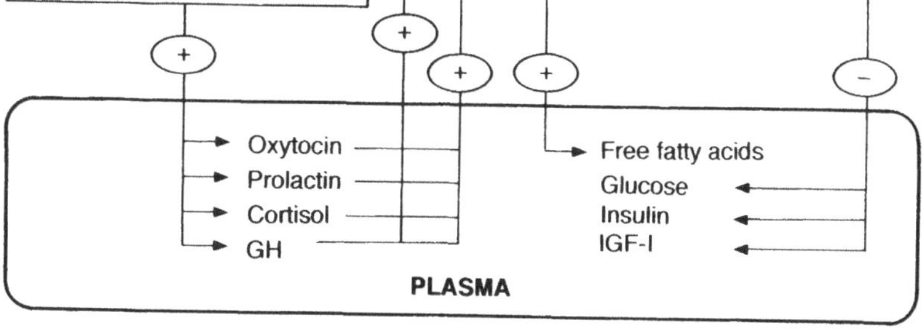

Fig. 6.2. Diagrammatic representation of neuroendocrine and metabolic factors involved in suckling and milk production in the sow (from Quesnel and Prunier, 1995).

Almond and Dial (1990) in the USA indicated that sows failing to return in oestrus after weaning had LH and oestrous responses to exogenous oestrogens similar to those of sows that had resumed cyclical ovarian activity. Such results indicated that the hypothalamic–pituitary axis of the anoestrous sow did not have a decreased sensitivity to the positive feedback effects of oestradiol. It is known that seasonal variations in response to exogenous oestrogens may occur in pigs (Cox *et al.*, 1987) and this may be a reason for the differing seasonal responses noted in weaned sows.

6.3.3. Nutritional factors

As well as studies dealing with endocrine factors influencing the rebreeding of pigs, there has been considerable research activity on the effect of nutrition on the duration of the interval between weaning and rebreeding. Delayed resumption of oestrus activity is known to be one of the main reasons for the premature culling of breeding animals. In Ireland, 30–40% of sows in Irish herds are culled annually, many of these for reproductive problems such as anoestrus. It is well established that low lactation feed intake can result in a longer weaning-to-oestrus interval (Reese *et al.*, 1984) and studies by Kirkwood *et al.* (1987) led them to conclude that low-plane feeding during lactation led to a significant reduction in the reproductive performance of pigs.

In Ireland, results of a study by Carroll *et al.* (1993) suggested that first parity sows with low feed intakes during lactation and after weaning suffered substantial weight loss and this was associated with an increase in the incidence of anoestrus after weaning; the authors suggested that such problems should be avoided by a feed regime that ensured high feed intakes during lactation, maximizing conservation of body condition during lactation and minimizing weight loss. In Canada, Zak *et al.* (1995) examined the effect of different patterns of nutrition during lactation in the primiparous sow; data indicated that the best response was achieved when sows were continually fed 'to appetite' during the 28-day lactation period.

According to Robinson (1990), evidence is now accumulating to show the importance of dietary protein in reducing the rebreeding interval. In Australia, for example, King and Martin (1989) demonstrated that LH concentrations in primiparous sows in the 20-h period prior to weaning were significantly higher in those receiving high (746g crude protein daily) rather than low (404g) protein levels; they suggested that the effect of protein intake during lactation on subsequent ovarian activity was mediated via effects on the secretion of LH pulses.

6.3.4. Stress effects

It is also evident, from various lines of research, that weaning may have a detrimental effect on the health status of the sow. There is not only the fact that separation of mother and offspring can affect the piglets, but weaning can influence the sow itself. Kyriakis *et al.* (1991) reported a consistent and significant decrease in the number of empty days in sows (from a mean of 15.0 to 11.7 days) treated with a psychotropic compound (amperozide). This compound has been shown to have antistress properties in pigs exposed to acute and chronic social stress (Bjork, 1989). From their study in Greece, it was concluded that behavioural factors could have a negative effect on the sow at weaning time, by delaying the return to oestrus. The authors took the view that factors other than those related to poor housing and management may adversely affect sow productivity and that amperozide reduced the emotional and biological responses of the animal to novel or threatening situations; controlling stress is likely to be important at time of weaning, as well as at farrowing. The compound amperozide has been shown to have antistress properties in pigs exposed to acute and chronic social stress.

6.3.5. Boar effect

Various workers have shown that certain environmental factors may influence the resumption of oestrus after weaning, including high energy diets, a change in environment (Rojanasthien, 1989) and the presence of the boar after weaning (Hemsworth, 1982). In Canada, Walton (1986) examined the effect of boar exposure during the last week of nursing and postweaning on the onset of oestrus and ovulation in primiparous and multiparous sows. They found that exposure of weaned sows to the boar immediately after weaning ensured a rapid return to oestrus and ovulation; boar presence during lactation was able to reduce the weaning-to-oestrus interval in older but not in first litter sows.

6.3.6. Effect of social environment

The effect of the sow's social environment at weaning on the onset of oestrus was investigated in Australia by Pearce and Pearce (1992). They showed that the exposure of weaned sows to an oestrous sow or boar would overcome the prolongation of the weaning-to-rebreeding interval which occurred during the summer and in primiparous sows.

6.3.7. Management of the lactating sow

Although the sow does not usually show heat periods during lactation, there are apparently certain circumstances in which it may be possible to overcome the inhibitory effect on ovarian activity. Two management practices have been mentioned in instances where oestrus has occurred during lactation; these involved either partial separation of the sow and litter or grouping of sows and their litters in the presence of a boar. Partial separation of sow and litters has proved effective in some instances but without effect in others.

A management system in which sows and their litters have been grouped in the presence of a boar and under an *ad libitum* feeding regime, was found by some to be effective in inducing oestrus in sows within 5 weeks of parturition (Rowlinson *et al.*, 1975). In this system, groups of three to eight sows and their litters were exposed to continuous physical boar exposure from day 24 postpartum to weaning at day 42. The interval from grouping to oestrus was 11 days; in one trial, all the lactating sows showed a heat period and, of those bred, 85% conceived.

In contrast to such findings, authors elsewhere in the UK recorded no appreciable incidence of oestrus. In attempting to analyse the reasons for apparent lack of success, Petchey and Jolly (1979) suggested that the experience of the sow in the previous pregnancy may have been a factor; failure in response may have been related to low feed levels or a negative nitrogen balance during the previous pregnancy.

Although recording a low incidence of oestrus in their groups of lactating sows, Petchey and English (1980) noted that the interval between exposure to the boar and the onset of the heat period was much shorter when the boar was absent from the group; it was thought that the boar probably had a crucial, although secondary, role to play in stimulating oestrus, possibly by influencing the sow's gonadotrophin status.

Ovulation frequency during late lactation among sows which were either group-housed or individually housed, without boar contact, was measured by Hulten *et al.* (1995) in Sweden. They applied a progesterone assay to faecal samples collected at weekly intervals from week 3 of lactation to 3 weeks after weaning. Although none of the individually housed pigs ovulated, 28% of those that were group-housed did so; the ovulation frequency of primiparous sows, sows in their second to fourth parity and sows in their fifth or greater parity was 0, 6 and 48%, respectively. Results showed the effect of social environment and age on ovulation frequency in the lactation period.

6.3.8. Oestrus in lactating Meishan pigs

In recent years, the Chinese Meishan breed of pig has aroused much scientific interest as a result of its remarkable reproductive capabilities; these include the fact that puberty can occur at about 3 months, average litter sizes are 30–40% larger than usual and there can be some incidence of fertile oestrus during lactation. From a study of Large White and Meishan sows in France, Elsaesser *et al.* (1992) concluded that there may be minor breed differences in the control of LH secretion and the functioning of the positive oestrogen feedback mechanism during lactation; there were indications that blockade of oestrogen feedback on LH release during lactational anoestrus may be less rigid in Meishan sows.

6.3.9. Early weaning and rebreeding

In normal commercial practice, piglets are weaned from the sow at a wide range of ages, including the usual 5-week weaning system down to very early weaning at 10 days of age. During recent decades, pig producers in many countries have tended to reduce the time the sow spends in lactation in pursuit of greater pig productivity; it was recognized that earlier rebreeding of sows after farrowing was one of the few areas of pig reproduction in which increases in output were still possible. In the UK, Dial and Bevier (1988) noted the increase in the percentage of herds weaning between 19 and 35 days after farrowing in the decade ending in 1984. They also draw attention to the inverse relationship between lactation length and the interval from weaning to oestrus. They estimated that for each 10-day reduction in lactation length, there was an increase of about 1 day in the weaning-to-conception interval.

6.3.10. Decline in litter size

Whether it was attempted at an experimental station or in commercial practice, early weaning, particularly with piglets below 21 days of age, was often accompanied by a decrease in the size of the subsequent litter. Cole *et al.* (1975) examined the effect of lactations varying in length from 4 to 42 days on subsequent litter size and recorded an average decrease from 12.7 piglets per litter at 21 days to 9.6 when the young were removed earlier; it was believed that such reductions were the result of a higher than usual incidence of embryo mortality in the early weeks of pregnancy (Varley and Cole, 1976).

From data generated in the UK and elsewhere, the indications were that weaning at 7 days of age or earlier was liable to reduce subsequent litter size by an average of two piglets. Newport (1977) calculated that this reduced potential sow productivity after very early weaning to little more than that expected with weaning at 21 days. Experimental work had also shown several adverse aspects of very early weaning, other than reduced litter size; these included an increase in the duration and variability of the interval between weaning and first fertile oestrus, reduced conception rate and an increase in the incidence of cystic follicles. In first parity sows, early weaning was found to pose a particular difficulty because of the prolonged weaning-to-oestrus interval which often occurred and the disappointing average size of second litters.

6.3.11. PG600 treatment

The effects of early (5–11 days) and later (23–31 days) weaning on the response of sows to a commercially available gonadotrophin treatment (PG600 = 400IU PMSG + 200IU hCG) was reported by Tonn *et al.* (1995) in the USA. Sows that were weaned early had significantly lower ovulation rates (15.9

vs. 24.0) and fewer embryos at 11 days (9.9 vs. 17.4) than those weaned later. The PG600 treatment was significantly more effective in reducing the interval to oestrus for early- than for later-weaned sows.

6.3.12. Intrauterine infusions in postfarrowing sows

Some authors have investigated the effect of intrauterine infusions on the subsequent reproductive performance of sows. In Japan, Moriyoshi *et al.* (1994) treated sows within 8 days of farrowing with a 100 ml intrauterine infusion of a 2% solution of polyvinylpyrrolidone-iodine; litter size was significantly increased over that in controls in sows that were weaned at 3 weeks.

In summing up views on early weaning systems, it is apparent that a lactation length of 3 weeks should provide maximum sow output with minimum reproductive malfunction. There is considerable evidence to show that commercial sow productivity does increase when piglets are weaned at 3–4 weeks rather than at 5–8 weeks.

6.3.13. Split-weaning management systems

There have been reports on a system in which certain of the piglets are separated from the mother while the remainder stay with her. In Canada, for example, Gilbertson *et al.* (1989) removed the heaviest piglets at 3 weeks and the remainder at 4 weeks. They concluded that in the absence of excessive sow weight loss or prolonged weaning-to-oestrus intervals, introduction of split-weaning did not improve sow or litter performance.

6.3.14. Zero-weaning and its problems

The ultimate early-weaning system, zero-weaning, is that in which the piglets are removed from the sow at farrowing or within a few hours of birth and reared artificially. At Cambridge, Polge (1972) drew attention to the lack of information on the reproductive consequences of this system. Although he noted that the postpartum uterus was probably capable of functioning reasonably well within 2 weeks of farrowing, as uterine regeneration would be well advanced by then, he also observed that such uterine repair may occur more slowly than expected in the zero- or very early-weaned sow.

In Canada, Elliot *et al.* (1980), in a comparison with sows weaned at 30 days, found that zero-weaning resulted in an increase in the interval between farrowing and conception, a corresponding increase in the farrowing interval, a greater incidence of cystic follicles and a reduced conception rate. The same investigators administered gonadotrophins (PMSG and hCG) 48h post-partum to zero-weaned sows in an attempt to induce ovulation and establish normal cyclical breeding activity; this treatment offered no advantage over weaning at 3 weeks. The Canadian workers suggested that it might be worth trying to suppress pituitary function for a period postpartum to mimic the effect operating in the lactating sow. This, in fact, had been attempted in some earlier studies, using methallibure, a known inhibitor of pituitary activity.

6.3.15. *Suppressing early pituitary activity*

Polge (1972) drew attention to the fact that the development of cystic follicles and associated infertility could be a serious problem in systems of very early weaning and suggested that it might be worth suppressing pituitary gonadotrophin activity for a short period after farrowing. Walker and Eddie (1974) used the central inhibitor, methallibure, over a 12-day period after farrowing and administered PMSG on the last day of treatment; sows that received the inhibitor had substantially larger litters than did controls (11.7 vs. 8.5 piglets). In a further study, Walker and Eddie (1975) treated sows weaned 10 days after farrowing with methallibure for 5 days and with PMSG on the following day. Although sychronization and conception rate appeared acceptable, litter size was lower than normal. The authors concluded that the more prolonged treatment with the central inhibitor in their earlier study might have been preferable. Such findings were in fact not applicable commercially because of the ban on methallibure.

It was clear from the various reports that zero-weaning had an adverse effect on the subsequent reproductive performance of the sow and that there was no advantage, in terms of sow productivity, over a system of 3-week weaning. Elliot *et al.* (1980) found that the only feasible means of overcoming the reproductive shortcomings associated with zero-weaning would be by combining the practice with the production of the once-bred gilt, a system in which the gilt is marketed soon after weaning.

6.4. Artificial Induction of Oestrus

As already mentioned, there is evidence to suggest that lactational anoestrus in pigs is the result of the suckling stimulus acting via the hypothalamus to inhibit the synthesis and release of LH and probably FSH secretion. Attempts that have been made over the years to induce oestrus and ovulation in the lactating sow have generally centred on overcoming this gonadotrophin deficiency, either by attempting to facilitate the release of the animal's own gonadotrophins or by using exogenous LH/FSH preparations.

Work in the 1940s and 1950s showed that an ovulatory oestrus may be induced in lactating sows by the administration of gonadotrophic hormones. The injection of an appropriate dose of PMSG (100–2000 IU), given after day 40 of lactation was shown to be consistently effective in inducing 86–96% sows to show oestrus 4 or 5 days later (Cole and Hughes, 1946; Heitman and Cole,

1956). When given at an earlier stage of lactation (e.g. day 20), the response proved to be both lower and much less consistent; in some instances, it was as low as 20% (Cole and Hughes, 1946) or 17% (Allen *et al.*, 1957). It also became clear that response to gonadotrophins could be influenced by genetic and management factors; Crighton (1970) employed a system of partial weaning in conjunction with PMSG at day 20 of lactation and found that this facilitated oestrous weaning (80% vs. 38%) as compared with results found using gonadotrophin alone.

Eastern European countries made extensive use of hormonal therapy in their pig-breeding programmes during the 1970s and 1980s. In the former East Germany, for example, Heinze and Stahl (1989) reported injecting 7062 sows with 1000IU PMSG 24h after weaning (at 5 weeks) in the winter/early spring period and a further 4616 in summer. Such treatments apparently came to an end with the reunification of Germany. A report from Russia (Nartya *et al.*, 1991) similarly dealt with the use of PMSG on the day of weaning as an aid to the induction of oestrus and ovulation.

6.4.1. Inducing pregnancy in lactating sows

There have been several reports of attempts to induce conception in lactating sows. There was a particularly effective protocol used by Hausler *et al.* (1980), in which PMSG/hCG treatment (1500IU PMSG with 1000IU hCG 96 h later) was followed by artificial insemination at a predetermined time (24 and 36/42h after hCG). Ovulation and conception could be induced as early as 15 days after farrowing without influencing the normal growth rate of the piglets. The conception rates, number of ovulations and litter sizes were apparently comparable to those achieved in a normal breeding programme. It was evident that under the particular animal and management conditions of their trial, the ovaries of lactating sows were responsive in some instances as early as day 10 and certainly by day 15; conception failures in sows treated at day 10 were thought to result from inadequacies in the uterine environment.

Cox and Britt (1982a) suggested that the initiation of oestrus and ovulation after weaning may be associated with changes in the secretion and release of GnRH. Studies reported by Clark *et al.* (1986), in which GnRH was given a week before weaning, did not show the treatment to be beneficial. However, earlier work by Cox and Britt (1982b) had shown that oestrus and ovulation in lactating sows could be induced by an appropriate administration procedure (hourly injections of GnRH).

At Aberdeen, Hatet *et al.* (1994) used the combined stimuli of sow grouping, high plane of nutrition and the presence of a mature boar in an attempt to induce oestrus in lactating sows, but these management methods were only partially successful in the induction of a synchronized oestrus. Earlier work elsewhere had shown that administration of exogenous gonadotrophin to recently weaned sows would shorten and reduce variation in the weaning-to-oestrus interval (Lancaster *et al.*, 1985) and that the efficacy of such treatment varied with stage of lactation (Britt *et al.*, 1985). In an examination of the role of stage of lactation (2 or 3 weeks) and gonadotrophin (PG600) at Aberdeen, Edwards *et al.* (1995) recorded that regardless of lactational stage, oestrus could be advanced and synchronized by PG600 but this was associated with a significant decrease in conception rate. The Aberdeen workers concluded that concurrent pregnancy and lactation cannot yet be reliably and synchronously established in grouped sows under commercial conditions.

6.5. References

Allen, A.D., Lasley, J.F. and Uren, A.W. (1957) The effects of gonadotrophic hormone injections on induction of estrus in lactating sows. *Journal of Animal Science* 16, 1097–1098.

Almond, G.W. and Dial, G.D. (1990) Sensitivity of the hypothalamo-hypophyseal axis to the positive feedback effects of estrogen in ovariectomized anestrous and diestrous sows. *Animal Reproduction Science* 22, 243–252.

Baker, L.N., Woehling, H.L., Casida, L.E. and Grummer, R.H. (1953) Occurrence of estrus in sows following parturition. *Journal of Animal Science* 12, 33–38.

Bjork, A. (1989) Is social stress in pigs a detrimental factor to health and growth that can be avoided by amperozide treatment? *Applied Animal Behaviour Science* 23, 39–47.

Blair, R.M., Nichols, D.A. and Davis, D.L. (1994) Electronic animal identification for controlling feed delivery and detecting estrus in gilts and sows in outside pens. *Journal of Animal Science* 72, 891–898.

Britt, J.H., Szarek, V.E. and Levis, D.G. (1983) Characterization of summer infertility of sows in large confinement units. *Theriogenology* 20, 133–140.

Britt, J.H., Armstrong, J.D., Cox, N.M. and Esbenshade, K. (1985) Control of follicular development during and after lactation in the sow. *Journal of Reproduction and Fertility* Suppl. 33, 37–54.

Carroll, C.M., Lynch, P.B., Boland, M.P., Austin, F.H. and Roche, J.F. (1993) The effect of feed intake on the reproductive performance of first parity sows. *Proceedings of the British Society Animal Production* (Winter Meeting), Paper no. 131.

Clark, J.R., Komkov, A. and Tribble, L.F. (1986) Effects of parity, season, gonadotropin releasing hormone and altered suckling intensity on the interval to rebreeeding in sows. *Theriogenology* 26, 299–308.

Cole, H.H. and Hughes, E.H. (1946) Induction of estrus in lactating sows with equine gonadotrophin. *Journal of Animal Science* 5, 25–29.

Cole, D.J.A., Barley, M.A. and Hughes, O.E. (1975) Studies in the sow reproduction. 2. The effect of lactation length on the subsequent reproductive performance of the sow. *Animal Production* 20, 401–406.

Cox, N.M. and Britt, J.H. (1982a) Relationships between endogenous gonadotropin-releasing hormone, gonadotropins and follicular development after weaning in sows. *Biology of Reproduction* 27, 70–78.

Cox, N.M. and Britt, J.H. (1982b) Induction of fertile estrus in lactating sows with hourly injections of gonadotropin releasing hormone. *Journal of Animal Science* 55 (Suppl. 1), 49.

Cox, N.M., Ramirez, J.L., Matamoros, I.A., Bennett, W.A. and Britt, J.H. (1987) Influence of season on estrous and luteinizing hormone responses to oestradiol benzoate in ovariectomized sows. *Theriogenology* 27, 395–405.

Crighton, D.B. (1970) The induction of pregnancy during lactation in the sow; the effects of a treatment imposed at 21 days of lactation. *Animal Production* 12, 611–617.

Dial, G.D. and Bevier, G.W. (1988) Noninfectious causes of reproductive failure in the female pig. In: Laing, J.A., Brindley Morgan, W.J. and Wagner, W.C. (eds) *Fertility and Infertility in Veterinary Practice*, 4th edn. Baillière Tindall, London, pp. 120–139.

Dusza, L. and Krzymowska, H. (1981) Plasma prolactin levels in the sow during pregnancy, parturition and early lactation. *Journal of Reproduction and Fertility* 61, 131–134.

Edwards, S. and Foxcroft, G.R. (1983) Endocrine changes in sows weaned at two stages of lactation. *Journal of Reproduction and Fertility* 67, 161–172.

Edwards, S.A., Gall, K., Shanks, M. and Hamilton, H. (1995) The effect of grouping time and gonadotrophin treatment on lactational oestrus in multisuckled sows. *Proceedings of the British Society Animal Science* (Winter Meeting), Paper no. 30.

Elliot, J.I., King, G.J. and Robertson, H.A. (1980) Reproductive performance of the sow subsequent to weaning piglets at birth. *Canadian Journal of Animal Science* 60, 65–71.

Elsaesser, F. and Parvizi, N. (1980) Partial recovery of the stimulatory oestrogen feedback action on LH release during late lactation in the pig. *Journal of Reproduction and Fertility* 59, 63–67.

Elsaesser, F., Martinat-Botte, F., Parvizi, N. and Terqui, M. (1992) Oestrus and LH responses to oestradiol during lactational anoestrus in Chinese Meishan and Large White sows. *Reproduction Nutrition Development* 32, 307–312.

Foxcroft, G.R. and Cosgrove, J.R. (1994) Integrated studies of nutrition–reproduction interactions: a pig paradigm. *Journal of Reproduction and Fertility*, Abstract Series no. 14, p. 3.

Gilbertson, J., Thacker, P.A. and Kirkwood, R.N. (1989) The influence of altered weaning management on piglet growth and sow reproductive performance. *Canadian Journal of Animal Science* 69, 33–37.

Hatet, G., Edwards, S.A., Gall, K. and Arey, D.S. (1994) Effect of three lactation housing systems on sow and piglet performance and behaviour. *Proceedings of the British Society of Animal Production* (Winter Meeting), Paper no. 189.

Hausler, C.L., Hodson, H.H.Jr, Kuo, D.C., Kinney, T.J., Rauwolf, V.A. and Strack, L.E. (1980) Induced ovulation and conception in lactating sows. *Journal of Animal Science* 50, 773–778.

Heinze, A. and Stahl, U. (1989) Investigation on the stimulation of oestrous cycle in the summer months in sows with synchronized ovulation, using different doses of PMSG. *Montshefte fur Veterinarmedizin* 44, 569–570.

Heitman, H.Jr and Cole, H.H. (1956) Further studies in the induction of estrus in lactating sows with equine gonadotropin. *Journal of Animal Science* 15, 970–977.

Hemsworth, P.H. (1982) Social environment and reproduction. In: Cole, D.J.A. and

Foxcroft, G.R. (eds) *Control of Pig Reproduction.* Butterworths, London, pp. 585–601.

Hulten, F., Dalin, A.M., Lundeheim, N. and Einarsson, S. (1995) Ovulation frequency among sows group-housed during late lactation. *Animal Reproduction Science* 39, 223–233.

Kann, G.D., Maretinet, J. and Schirar, A. (1976) Impairment of LH release following oestrogen administration to hyperprolactinaemic ewes. *Nature* 264, 465.

King, R.H. and Martin, G.B. (1989) Relationships between protein intake during lactation, LH levels and oestrous activity in first-litter sows. *Animal Reproduction Science* 19, 283–292.

Kirkwood, R.N., Baidoo, S.K., Aherne, F.X. and Sather, A.P. (1987) The influence of feeding level during lactation on the occurrence and endocrinology of the postweaning estrus in sows. *Canadian Journal of Animal Science* 67, 405–415.

Kyriakis, S.C., Olsson, N.G., Martinsson, K. and Bjork, A.K.K. (1991) Observations on the action of amperozide: are there social influences on sow-litter productivity? *Research in Veterinary Science* 51, 169–173.

Lancaster, R.T., Foxcroft, G.R., Boland, M.P., Edwards, S. and Gordon, I. (1985) Fertility of sows treated with exogenous estradiol and/or gonadotropins to control post weaning estrus. *Animal Reproduction Science* 8, 365–373.

Landeghem, A.A.J.Van and Wiel, D.F.M.Van de (1978) RIA for porcine prolactin: plasma levels during lactation, suckling and weaning after TRH administration. *Acta Endocrinologica* 88, 653–658.

Li, Y., Huang, C.J., Klindt, J. and Anderson, L.L. (1991) Divergent effects of antiprogesterone, RU 486, on progesterone, relaxin and prolactin secretion in pregnant and hysterectomized pigs. *Endocrinology* 124, 1294–1299.

Maurer, R.R., Ford, J.J. and Christenson, R.K. (1985) Interval to first postweaning estrus and causes for leaving the breeding herd in Large White, Landrace, Yorkshire and Chester White females after three parities. *Journal of Animal Science* 61, 1327–1334.

Moriyoshi, M., Kato, M., Hiroi, N., Nakao, T. and Kawata, K. (1994) Effect of intrauterine infusion of 2% PVP-I solution on the reproductive performance in post-farrowing sows, with comparisons of two farms with different length of suckling. *Japanese Journal of Swine Science* 31, 116–119.

Nartya, V.G., Korotkov, A.A. and Dmitriev, V.B. (1991) Stimulation of oestrus and ovulation in sows. *Zootekhniya* 10, 58–60.

Newport, M.J. (1977) Early weaning of pigs: a major advance in pig production. *World Animal Review* 24, 34–39.

Palmer, W.M., Teague, H.S. and Venzke, W.G. (1965) Histological changes in the reproductive tract of the sow during lactation and early post weaning. *Journal of Animal Science* 24, 1117–1125.

Parvizi, N., Elsaesser, F., Smidt, D. and Ellendorff, F. (1976) Plasma LH and progesterone in the adult female pig during the oestrous cycle, late pregnancy and lactation, and after ovariectomy and pentobarbitone treatment. *Journal of Endocrinology* 69, 193–203.

Paterson, A.M. and Pearce, G.P. (1994) Plasma hormone and metabolite concentrations and the interval from weaning to oestrus in primiparous sows. *Animal Reproduction Science* 36, 261–279.

Pearce, G.P. and Pearce, A.N. (1992) Contact with a sow in oestrus or a mature boar stimulates the onset of oestrus in weaned sows. *Veterinary Record* 130, 5–9.

Petchey, A.N. and English, P.R. (1979) Sow service in lactation: an analysis of data from one herd. *Animal Production* 29, 183–191.

Petchey, A.N. and English, P.R. (1980) A note on the effects of boar presence on the performance of sows and their litters when penned as groups in late lactation. *Animal Production* 31, 107–109.

Polge, C. (1972) Reproductive physiology in the pig with special reference to early weaning. *Proceedings of the British Society of Animal Production* 1, 5–18.

Quesnel, H. and Prunier, A. (1995) Endocrine bases of lactation anoestrus in the sow. *Reproduction, Nutrition, Development* 35, 395–414.

Ramirez, J.L., Cox, N.M. and Moore, A.B. (1994) Enhancement of litter size in sows by insulin adminstration prior to breeding. *Journal of Animal Science* 72 (Suppl. 1)/*Journal of Dairy Science* 77 (Suppl. 1), 79.

Reese, D.E., Peo, E.R. and Lewis, A.J. (1984) Relationship of lactation energy intake and occurrence of post weaning estrus to body and backfat composition in sows. *Journal of Animal Science* 58, 1236–1244.

Renesis, F.De, Hunter, M.G. and Foxcroft, G.R. (1993a) Suckling-induced inhibition of luteinizing hormone secretion and follicular development in the early postpartum sow. *Biology of Reproduction* 48, 964–969.

Renesis, F.De, Cosgrove, J.R. and Foxcroft, G.R. (1993b) Luteinizing hormone and prolactin responses to naxolone vary with stage of lactation in the sow. *Biology of Reproduction* 48, 970–976.

Robinson, J.J. (1990) Nutrition in the reproduction of farm animals. *Nutrition Research Reviews* 3, 253–276.

Rojanasthien, S. (1989) Effect of transportation and relocation in post-weaning anoestrous primiparous sows. *Acta Veterinaria Scandinavica* 30, 1–8.

Rowlinson, P., Boughton, H.G. and Bryant, M.J. (1975) Mating of sows during lactation: observations from a commercial unit. *Animal Production* 21, 233–241.

Sterning, M., Rydhmer, L., Einarsson, S. and Andersson, K. (1994) Oestrous symptoms in primirarous sows.1. Duration and intensity of external oestrous symptoms. *Animal Reproduction Science* 36, 305–314.

Stevenson, J.S. and Britt, H.H. (1981) Interval to estrus in sows and performance of pigs after alteration of litter size during late lactation. *Journal of Animal Science* 53, 177–181.

Svajgr, A.J., Hays, V.W., Cromwell, G.L. and Dutt, R.H. (1974) Effect of lactation duration and reproductive performance of sows. *Journal of Animal Science* 38, 100–105.

Tonn, S.A., Groothius, P.G., Boese, B.J., Blair, R.M. and Davis, D.L. (1995) Effects of vitamin A and gonadotropin treatments on estrus and early pregnancy for sows weaned at 5–11 (early) or 23–31 (late) d postpartum. *Journal of Animal Science* 73 (Suppl. 1), 91.

Varley, M.A. and Cole, D.J.A. (1976) Studies in sow reproduction. The effect of lactation length of the sow on the subsequent embryonic development. *Animal Production* 22, 79–85.

Varley, M.A. and Foxcroft, G.R. (1990) Endocrinology of the lactating and weaned sow. *Journal of Reproduction and Fertility* Suppl. 40, 47–61.

Walker, N. and Eddie, S.M. (1974) The effect on sow productivity of oestrus suppression following a two-day lactation. *Animal Production* 18, 153–158.

Walker, N. and Eddie, S.M. (1975) A note on the effect of sow productivity of methallibure administration following a 10-day lactation. *Animal Production* 20, 303–305.

Walton, J.S. (1986) Effect of boar presence before and after weaning on estrus and

ovulation in sows. *Journal of Animal Science* 62, 9–15.
Zak, L.J., Cosgrove, J.R., Aherne, F.X. and Foxcroft, G.R. (1995) Effect of feeding level during lactation upon reproductive function in the primiparous lactating sow. *Journal of Animal Science* 73 (Suppl. 1), 306.

7 Increasing Litter Size in Pigs

7.1. Introduction

Litter size in the pig, as in other farm animals, is a function of ovulation rate, fertilization rate and intrauterine mortality; obviously, the number of oocytes released from follicles at time of ovulation will set the upper limit of litter size for any particular gestation period. A number of authors have reviewed factors affecting ovulation rate and litter size as well as drawing attention to the way in which the uterus imposes a limitation upon the number of fetuses surviving in the pig. It may limit by virtue of its length (its actual inability to hold more than a certain number of young) and it may limit by the evenness or otherwise of the distribution and spacing of the embryos and fetuses (Dziuk, 1992).

Upper limits of litter size

Clark and Leman (1986) arranged the many factors known to influence litter size in the pig into two major categories: (1) those for which data are available from the records kept in commercial units (age, parity, number of matings per conception, season, previous lactation length, previous weaning-to-conception interval); and (2) those variables for which measurements were not feasible (effect of boar, husbandry practices, disease, stress). From research, it is possible to gather information that is not otherwise available. It is believed that litter size is related to the sow's ovulation rate until it reaches a total of 14 fetuses or seven fetuses per uterine horn (Wu *et al.*, 1989); litter size beyond 14 fetuses is apparently related to uterine length but not to number of ovulations (Wu *et al.*, 1987).

7.1.1. Factors influencing ovulation rate

Many authors have reported on factors influencing the ovulation rate in gilts and sows. Many of these factors relate to the animals themselves and to the housing and management conditions in which they are kept. Some effects are also due to environmental factors (daylength, humidity and temperature) over which the producer may have less control. In Spain, Falceto *et al.* (1993a) analysed data on the genital tracts of some 1200 sows and showed that although ovulation rate was not significantly affected by hours of daylight or air humidity, there was a negative correlation between ovulation rate and ambient temperature. This agrees with many reports showing the adverse effect of high temperature on reproductive function in the pig. A further report from Spain showed that ovarian activity was significantly affected by season, the ovulation rate being highest in April and September and lowest in summer (Falceto *et al.*, 1993b); the authors recorded a marked increase in ovulation rate during September and October.

It is evident that when the sow's ovulation rate is increased to well beyond the normal physiological limits (e.g. using exogenous gonadotrophins), the benefits of any small increase in the average litter size will almost invariably be outweighed by the variability in the size of the litters and in the birthweights of piglets in these litters. Certainly, there have been those who have concluded that although ovulation rate should be kept as high as possible by natural means, attempts to produce increases artificially would be of no commercial interest. Reviews of the literature dealing with factors influencing litter size in the sow are given in Clark (1986), Clark and Leman (1986) and Buttle (1991).

7.2. Factors Affecting Litter Size in Pigs

Reproductive performance is measured primarily by the number of living pigs at birth in the sow; under normal systems of weaning (i.e. 3 weeks and above), a target of 11–12 piglets born alive per litter should be the aim for the sow and 9–10 piglets for the gilt (Fig. 7.1). According to Clark (1986), there was no great evidence at that time to suggest that litter size at birth or weaning had improved over the years, despite the many management schemes that have been developed to improve the reproductive efficiency of the pig; in the USA, the number of pigs weaned per litter varied little from 7.4 in the 40 years up to the mid-1980s. Although there may be many causes of small litters, it is generally recognized that the problem is usually not a result of disease factors.

7.2.1. Breed effects

Ovulation rate in pigs is highly heritable and responds to selection (Johnson *et al.*, 1985). However, an increased ovulation rate is accompanied by a decrease in the embryo/fetal survival rate. The problem of increasing litter size would be much easier if ovulation rate and embryo/fetal survival could be improved simultaneously. Although it is clear that some breeds have higher litter sizes than others (e.g. Meishan), there is much evidence to show that litter size may vary within a breed just as much as between breeds. The heritability of litter

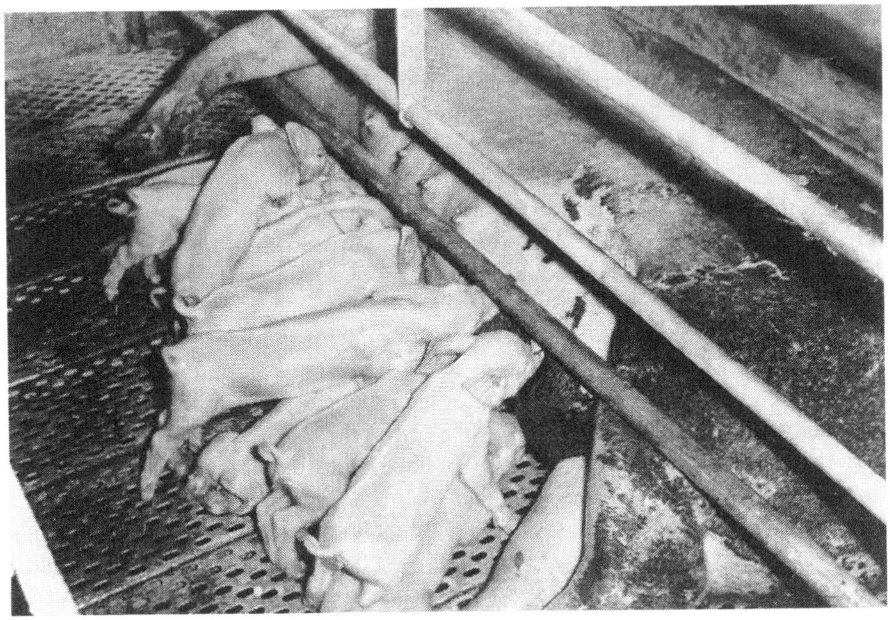

Fig. 7.1. A good litter size is essential for profitability in pig production.

size is known to be very low and to be controlled primarily by the genotype of the sow.

There have been those who maintain that there may be little to be gained in selecting for the trait; an extensive review of the literature in the 1970s showed heritability to be no more than about 10%. On the other hand, pig crossbreeding studies, dating from the 1930s, using inbred and other animals, have shown that heterosis for reproductive traits, including litter size, ranges from 5% to 25%, depending on the breeds used in the cross (Johnson *et al.*, 1978). Consequently, it has become a widespread commercial practice in many countries to use crossbred or hybrid sows as the breeding females in commercial units.

Strategies that have been developed to overcome the problem of low heritability have been reviewed by Webb (1995); these include the group nucleus breeding scheme which uses best linear unbiased prediction (BLUP) and artificial insemination to connect together several herds to provide a large nucleus population (Fig. 7.2). According to Webb (1995), the Cotswold Group Nucleus Breeding Scheme is expected to permit litter size to increase by 0.2 live piglets per litter annually. Although theoretically optimal, selection strategies based on the measurement of ovulation rate, embryo survival and uterine capacity by laparotomy would probably be neither economically worthwhile nor acceptable on welfare grounds.

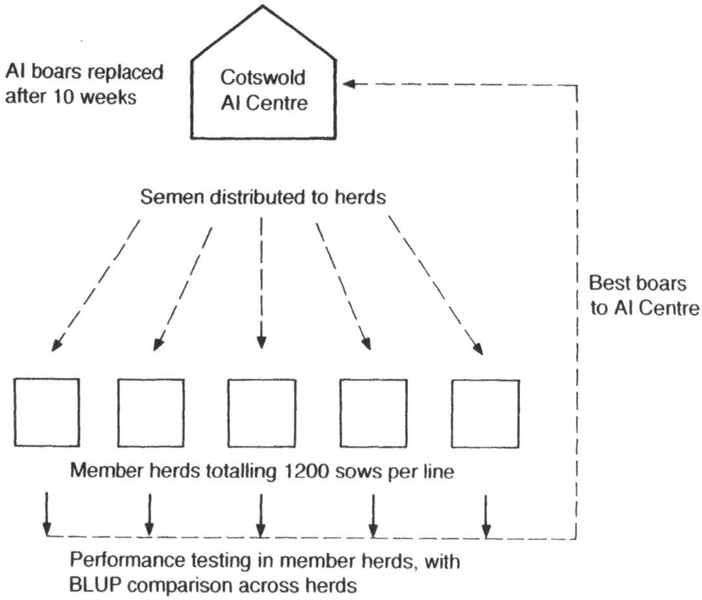

Fig. 7.2. The Cotswold Group Nucleus Breeding Scheme for pigs (from Webb, 1995).

7.2.2. The Meishan breed and litter size

It is clear from many reports that the Chinese Meishan is one of the most prolific breeds in the world, producing three to four more piglets per litter than sows of the European breeds. Although the pure bred Meishan is not commercially viable in the UK or Ireland because of its poor growth and carcass characteristics, it is being used in the formation of certain synthetic pig lines. The purebred Meishan is of considerable interest, however, to those researchers who are attempting to understand the physiological and endocrinological factors responsible for variations in litter size.

In the USA, Bazer *et al.* (1988) suggested that faster and more uniform conceptus development between days 8 and 14 might account for higher embryonic survival rates in the Meishan; not all workers agreed that this was the only mechanism involved. If early embryonic loss was due to asynchrony of some embryos and the uterine environment, it might be expected that smaller or slower developing individuals would fail because the advancing uterine environment favoured the larger and more differentiated blastocysts. In Canada, however, Cassar *et al.* (1994) showed that most male pig embryos grew faster than females before they became attached and yet the sex ratio was essentially 1:1; they concluded that it was unlikely that embryo survival was either sex- or size-dependent.

It is evident that the Meishan in the UK has a significantly greater ovulation rate than do Large White pigs of a similar age (Galvin *et al.*, 1993). It is also apparent that Meishan sows have a greater 'uterine capacity' and for that reason are capable of maintaining a higher number of fetuses than is possible with other breeds. There is apparently an ability of the Meishan female to produce smaller fetuses and piglets than sows of European breeds; maternal control of fetal growth may be an important factor in explaining the high litter sizes that can be maintained. A further factor favouring the Meishan may be the better control of spacing between fetuses in the uterus; it is not clear whether this is due to more even attachment or because of differential growth of the uterus after attachment.

7.2.3. Marker-assisted selection for litter size

The oestrogen receptor (ESR) gene has been shown to be significantly associated with litter size in a 50% Meishan synthetic pig line. Short *et al.* (1995) showed that each copy of the favourable marker is associated with an additional 1.0 piglet born alive per litter in this synthetic line. These workers report the use of the ESR genotype to increase the accuracy of selection for litter size in commercial dam lines.

7.2.4. Nutritional effects

Adequate nutrition of the gilt and sow is obviously important; the ovulation rate can be influenced by feeding before oestrus and evidence suggests that total feed or energy intake is of greatest significance. Although a deficient nutritional regime may depress ovulation rate (Brooks and Cooper, 1972), there is less certainty about the way in which nutrition can be a factor influencing ovulation in the sow. In this regard, the pig differs from the sheep, where it is less easy to show evidence of nutritional effects in ewe lambs than in older animals. There is ample evidence to show that 'flushing' regimes applied during the follicular phase of the gilt's oestrous cycle almost always increase the ovulation rate (Cox *et al.*, 1987; Flowers *et al.*, 1989; Rhodes *et al.*, 1991), whereas in the sow such evidence is not found.

At one time, on the basis of work at Nottingham, it was thought that increased feed intake during oestrus might affect ovulation rate (Lodge and Hardy, 1968). There was a trend towards higher ovulation rates as postmating feed intake of sows was raised from 1.4 to 4.1 kg day^{-1}, although the differences were not significant. The effect on litter size of doubling the feed intake of sows on the day of mating was subsequently investigated over a wide range of management conditions in a field trial (Brooks and Cole, 1972) and it became evident that only in isolated instances did such short-term feeding produce an effect.

Elsewhere, later studies reported by various authors suggested that, after the birth of the first litter, large differences in feeding levels during lactation, leading to marked differences in bodyweight had no effect on subsequent ovulation rates (King and Dunkin, 1986; Baidoo and Aherne, 1988); high-plane feeding between weaning and rebreeding was also apparently without effect.

Over the longer term, it should be mentioned that in planning for a potential herd life of six litters, it was suggested that a sow should increase her liveweight by 12–15kg between successive matings at least up to the fourth litter (Brooks, 1976); such increases are the result of the anabolic effects of pregnancy in the pig.

7.2.5. Endocrinology of nutritional flushing of gilts

The endocrinology of nutritional flushing in gilts was the subject of a study by Beltranena *et al.* (1991) in Canada. They found evidence suggesting an increased frequency of episodic LH release; flushed gilts had a greater plasma insulin concentration around the time of oestrus and a greater insulin-like growth factor-I (IGF-I) concentration during the LH-surge period than control pigs. The authors considered that increased plasma insulin may be an important metabolic cue, which, in an independent or synergistic action with gonadotrophins and IGF-I, may mediate the ovulatory reponse to flushing in the gilt.

7.2.6. Parity and age effects

Age as a factor affecting litter size is well enough accepted, but there have sometimes been differences of opinion as to the source of the variations that may occur with age. Clark and Leman (1986) advised caution when interpreting the influence of parity on litter size because of the wide range of possible sow ages within a parity. Anderson and Melampy (1972) found that the number of piglets farrowed increased between the first and fourth litters, but by the eighth litter the number of stillborn young increased; when litter size was related to age, a decline was evident after about 4.5 years. French *et al.* (1979) discussed the view that previous reproductive 'experience' could increase litter size in the sow. The suggestion was advanced by some that such 'experience' on the part of the uterus could better prepare it for carrying a large number of young to term.

French *et al.* (1979) also drew attention to the fact that much data in the literature relating age and parity to litter size showed the two factors to be highly confounded; experimental evidence indicated that litter size continued to rise as the age of the sow increased (until at least 2 years of age) and parity had little or no direct effect on litter size. A large amount of information on the influence of parity on litter size was reviewed by Clark and Leman (1986), who noted that litter size was lowest in the first parity, reached maximum size in parities three, four and five and then remained nearly constant or slowly declined as parity increased.

7.2.7. Effect of delayed breeding

There is a particular problem in first parity sows returning to oestrus after weaning. Clark (1986) quotes a figure of almost 40% failing to return to oestrus within 2 weeks, in comparison with a figure of about 20% for sows of other parities. Those primiparous sows that do show oestrus may show a much reduced litter size (below herd average). If the those pigs that fail to show first oestrus are rebred at their next oestrus, however, they may produce large litters (above herd average). In Alberta, Clowes *et al.* (1994) examined the effect of delayed breeding on litter size in Cambrough sows; they concluded that delaying the breeding of first and second parity sows (to their second oestrus after weaning) significantly increased litter size, possibly by influencing embryo survival.

7.2.8. Sire effects and breeding procedures

Significant sire effects on the outcome of pregnancy have been recorded both in sheep and cattle, presumably the result of differences in the viability of embryos derived from certain males. The same is true of pigs (Rahnefeld and Swierstra, 1970). There is, of course, the possibility that differences could arise from boar usage. According to Dial and Bevier (1988), litter size should not be adversely affected if boars are used less than six times weekly; litter size is apparently less sensitive to boar overusage than farrowing rate.

According to Clark and Leman (1986), a single mating per oestrus performed at the optimum time during oestrus would be expected to achieve maximum litter size. Some authors measured the electrical conductivity of vaginal mucosa in attempts to determine the optimum breeding time. Hooper *et al.* (1982) reported on the use of the Walsmetta device to predict optimal times. They concluded that one natural mating timed with the Walsmetta was equal to two natural matings. In Ireland, O'Grady *et al.* (1983) reported similar findings.

Several reports have shown that two rather than one mating per oestrus can significantly increase litter size both in gilts and sows. Tilton and Cole (1982) found an increase of 1.3 pigs per litter when sows mated naturally three times at 24h intervals compared with pigs naturally mated twice at 24h intervals. Using AI, Reed (1982) similarly reported an increased litter size when three rather than two inseminations were compared. Although the general consensus in the literature has been that there is a positive association between the number of matings and litter size, there have still been those who take the view that a single mating is to be preferred.

7.2.9. Effect of lactation length

The results in many reports reviewed by Clark and Leman (1986) showed that a reduction of weaning age from day 42 towards day 0 was associated with a corresponding reduction in litter size. It was apparent from the literature that much of this reduction was associated with a decrease in embryonic survival rate. According to Varley (1982), the progesterone and oestradiol levels that follow short lactations may be a factor influencing the extent of embryonic mortality. There are certainly circumstances in which exogenously administered steroids may influence embryonic survival.

7.2.10. Effect of light and photoperiodicity

In the wild pig, reproduction is clearly seasonal with a biphasic pattern (Claus and Weiler, 1985). Although the main breeding season falls in late autumn, a second one may occur around April. As reported by Mauget (1982), from July to September, all sows are likely to be in anoestrus. In the domestic pig, sensitivity to photoperiodic stimuli has not entirely disappeared and, for this reason, there is some interest in considering seasonal effects and how light regimes may influence litter size. It is evident that litter size may be increased after mating in autumn compared with summer (Dobao *et al.*, 1983; Noguera *et al.*, 1983). Litter size from matings in summer may be one piglet less than those occurring in the autumn/winter period (Mattioli *et al.*, 1983).

Light programmes which extended the daily light period to a constant 15–16h apparently proved ineffective in improving the reproductive characteristics of sows; the simulation, in summer, of the decreasing photoperiod normally associated with autumn (20 min decrease per week) from May to August removed the seasonal increase of the weaning-to-oestrus interval (Claus and Weiler, 1985).

7.3. Hormonal Methods of Increasing Ovulation Rate

Although the injection of gonadotrophins can be used to increase ovulation rate in the pig (superovulation) for the purposes of embryo transfer or associated techniques (see Chapter 8), for normal commercial purposes, there is nothing to be gained in considering such an approach to increasing litter size. Apart from the use of gonadotrophins, studies have been reported on the use of altrenogest, on immunological approaches and other novel methods to an increase in ovulation rate.

7.3.1. Use of gonadotrophins

Anderson and Melampy (1972), reviewing data from many different reports, calculated that the injection of PMSG (doses of 500–1500IU) induced the release of an average of 4.8 addional oocytes, but that only one extra fetus survived at day 30 of gestation. There may, however, be some circumstances in which ovulation rate may be increased by hormonal means which deserve mention.

7.3.2. Use of altrenogest (Regumate)

It has been reported that the progestagen, altrenogest, when given orally (15 mg day^{-1}) for 2 weeks to gilts tended to increase the ovulation rate (Rhodes *et al.*, 1991). In the UK, Baker *et al.* (1994) administred altrenogest to primiparous sows in feed at 20 mg per head daily for 5 days, starting on day 2 after weaning; such treatment resulted in a significant increase in piglets born alive per litter (+ 0.9).

7.3.3. Immunization against inhibin and follistatin

Ovulation rate in the pig, as in other farm animals, is a complex process involving both extragonadal (e.g. FSH and LH) and intragonadal regulators (steroids and peptides such as inhibin, activin and follistatin). There has been much research activity in the development of inhibin-based fecundity vaccines for use in sheep and cattle in the past decade but only limited work has been reported for pigs. In Australia, for example, Brown *et al.* (1990) reported that immunization against recombinant bovine inhibin alpha subunit resulted in a significant increase in ovulation rates in gilts (15.9 vs. 12.2). They recorded no deleterious effect on growth rate or onset of oestrus of the animals. In the USA, King *et al.* (1993) also reported on responses in gilts after active immunization against a synthetic fragment of bovine inhibin; the treatment significantly increased ovulation rate by 39% (17.8 vs. 12.8). The authors found that FSH level was higher in the immunized gilts than in controls before the preovulatory surge of LH and lower after the surge.

Follistatin is a monomeric glycoprotein isolated from pig follicular fluid which is believed to be involved in the suppression of FSH as opposed to the synthesis of FSH (Tonetta and diZerega, 1990). Activin is known to influence gonadotrophin-stimulated events (Hutchinson *et al.*, 1987); this peptide can enhance FSH-stimulated induction of LH receptors. Follistatin apparently affects follicle growth by inhibiting the action of activin which stimulates folliculogenesis. Christensen *et al.* (1994) in Canada found that immunoneutralization of follistatin may be a means of increasing ovulation rate, and consequently litter size, in the gilt. These workers developed a cell-line which produced immunogenic follistatin; they reported that animal trials were under way to determine the physiological effect of neutralizing antibodies to follistatin and to develop a fecundity vaccine.

7.3.4. Use of Epostane

It has been shown that Epostane, an agent that prevents progesterone synthesis by inhibiting the activity of the enzyme 3β-hydroxysteroid dehydrogenase can increase ovulation rate in the pig. In the USA, Fu *et al.* (1990) reported significantly higher ovulation rates in gilts treated for either 7 or 12 days with

Epostane at two centres (21 vs.16 at one centre; 17 vs. 12 at the other). The influence of 7- and 12-day Epostane treatment on ovulation rate after day 14 of the gilt's oestrous cycle was similar, indicating that the important effect of the agent occurred after day 14. The number of viable embryos in gilts slaughtered 30 days after breeding was not significantly increased by the treatment.

7.4. Embryonic and Fetal Mortality in the Pig

From a research viewpoint, the question of litter size is one of examining the factors that influence the basic ovulation rate and, more important still, it involves an examination of factors affecting the incidence of embryonic mortality (Fig. 7.3). The sow differs from the other farm animals in that litter size depends less on the number of oocytes released from the ovaries and much more on the number of fetuses that survive in the uterus (Hammond, 1921). There is ample evidence to show that the incidence of prenatal mortality in the pig is greatest during early gestation (Pomeroy, 1960), with estimates of losses running as high as 40% or so.

7.4.1. The early pig embryo

From studies conducted to determine the time at which embryonic mortality occurs in the pig, Varley and Cole (1976) concluded that the bulk of loss takes

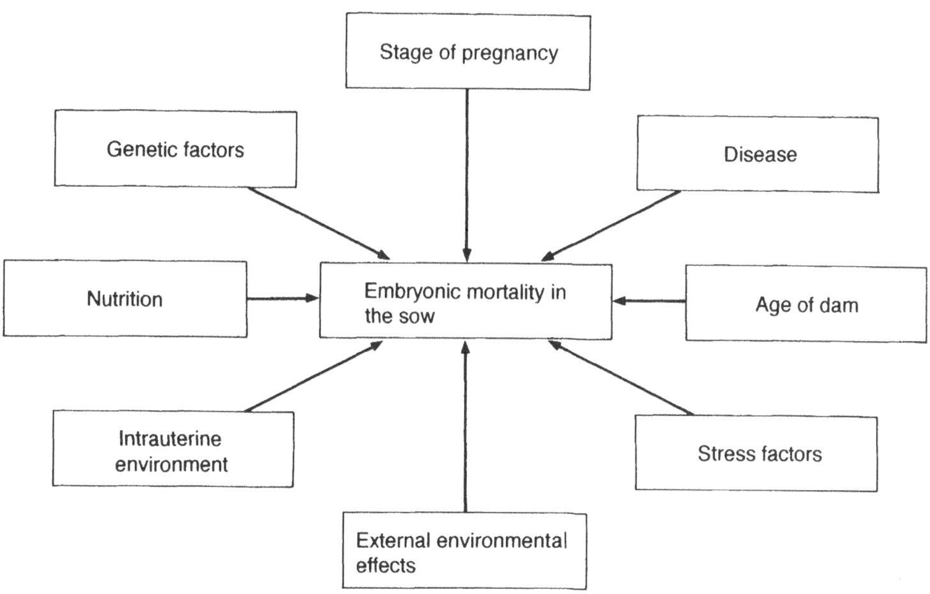

Fig. 7.3. Factors affecting embryonic mortality in the sow.

place at or around the time of attachment (i.e. in embryos aged 2–3 weeks). It is at this time that the pig embryo is undergoing marked elongation. At day 10 of pregnancy, the blastocyst is spherical measuring some 3–5 mm in diameter (Anderson, 1978); at 12 days, as a result of the rapid elongation of the trophoblast, it has increased to several centimetres in length.

7.4.2. Occupation of the uterine horns

By day 14, each embryo can occupy about 20–30 cm of uterus, each horn of which can be about 100–150 cm in length. Provided the number of embryos is sufficient, the whole uterine lumen is occupied by the elongating blastocysts. Points of loose contact between trophoblast and endometrium are evident at day 14, although intimate attachment is not established until day 18 (Crombie, 1970). The physiological and endocrinological mechanisms which influence uterine events in the early weeks of pregnancy have been the subject of intense research activity and much is now understood about the maternal recognition of pregnancy in the pig (see Chapter 4).

7.5. Attempts to Increase Embryo Survival

The extent of embryonic mortality in the sow has prompted various investigations into the effect on embryo survival of treating the sow with hormones during the early weeks of pregnancy. Workers at North Carolina, for example, reported that uterine capacity might be influenced by the level of progesterone around day 13 of gestation (Rampacek *et al.*, 1975). Although the administration of progesterone alone, even in very high doses, did not influence litter size, there was some evidence that a combination of progesterone and oestrogen might be more effective.

Morgan (1987) found that smaller blastocysts within a given litter may be indirectly destroyed by their advanced littermates, because of the ability of the larger embryos to alter the uterine environment. Pope (1992) discussed how subtle differences between littermate embryos may result in the death of some embryos. Oestrogen secretion by the larger blastocysts within a litter may influence the uterine environment to their advantage. It is clear that there is a very delicate intrauterine hormonal balance in the early weeks of pregnancy in the pig (Ashworth, 1991).

7.5.1. Progesterone : oestrogen treatments

Some years ago, there were indications that administration of progesterone and oestrogen (oestrone or oestradiol) in a 2000:1 ratio could result in an increase in the number of live embryos in gilts examined at midpregnancy (Day *et al.*, 1963). Results reported by Wildt *et al.* (1976) indicated that treatment with a

2000:1 progesterone:oestrone combination in early pregnancy (days 14 to 23 of gestation) had a beneficial effect on embryo survival. Morcom *et al.* (1976) similarly employed this hormome combination on days 16 and 17 and reported a tendency for the treated pigs to have larger litters; it was speculated that such treatment may have induced an increase in the uterine proteins usually synthesized in limited quantities during this critical phase of attachment. It is believed that such proteins, together with those secreted by the embryo, are part of the crucial interaction which must occur between dam and embryo if a successful pregnancy is to be established (Mirando *et al.*, 1990). Experiments reported by deSa *et al.* (1981) confirmed earlier findings on the beneficial effect of oestrone–progesterone treatment and showed that this could be achieved by administering the hormone combination on days 16 and 17, a 2-day period rather than an 8-day one.

Workers in the former East Germany, on the other hand, used progesterone alone or in combination with oestradiol in early pregnancy in gilts and sows without observing any positive effect (Heidler *et al.*, 1979). In the USA, although McGovern *et al.* (1981) found the progesterone–oestrone combination had no effect on litter size, they did record a transient effect on the development of the allantochorion membrane, which may confer a survival benefit under some conditions.

In Ireland, studies reported by Sheridan *et al.* (1986) indicated that the administration of oestrone and progesterone to sows and gilts around the time of attachment (days 16 and 17) had a benefical effect on the number of piglets born alive. When the combination treatment was confined to a single day (day 16 or day 17), there was a non-significant trend towards increased litter size.

In Wisconsin, Pope *et al.* (1987) examined the possibility that exogenous oestradiol alone might be a suitable treatment to improve litter size in a typical herd; although their results did not indicate that oestradiol treatment (2 mg oestradiol per day on days 12 and 13 of gestation) increased litter size, fewer animals returned to oestrus after breeding, suggesting that the treatment might be beneficial in maintaining events in early pregnancy. On the debit side, however, some sows produced significantly smaller litters than controls. In Florida, an examination of the hormonal regulation of IGF-I levels in the uterus in early pregnancy indicated that oestradiol and progesterone were involved in the normal regulation of uterine IGF-I synthesis and/or secretion (Simmen *et al.*, 1990).

A study by Knox and Zimmerman (1995) in Nebraska examined the possibility that treatment with FSH-P might be used to induce synchronized superovulation in pigs so as to achieve decreased variation in the stages of development exhibited by day 10 embryos. This work stemmed from reports indicating that embryonic loss in pigs may be due, in part, to asynchronous development during early gestation. Despite increased numbers of embryos in the FSH-P-treated gilts, they found no evidence of greater within-litter variation in embryo size; embryos sizes were apparently more synchronous than in PMSG-injected or control gilts.

7.5.2. Nutrition during pregnancy

During pregnancy, large variations in protein and energy intake have been shown to have profound effects on sow weight, but in the majority of cases have not resulted in significant differences in the numbers of piglets born. Results from various studies, on the other hand, have shown that high-plane feeding in early pregnancy can cause an increase in embryo mortality in gilts (Kirkwood and Thacker, 1988) but not in sows (Toplis *et al.*, 1983). There has also been the demonstration by Ashworth (1990, 1991) in gilts that decreased embryo survival arising from high-plane feeding could be avoided by the use of exogenous progesterone.

It is evident that in the pig, as in the ewe, high food intakes, via their stimulatory effects on hepatic blood flow and the metabolic clearance rate of progesterone (Symonds and Prime, 1989), may result in a decline in progesterone levels to the point at which embryo development and survival may be compromised. Aherne and Kirkwood (1985) had earlier noted that nutritional regimens may alter rates of steroid clearance by increasing hepatic blood flow and/or by enzyme activity in the liver. The same authors speculated that this mechanism might be implicated in trials in which high levels of feeding in early pregnancy were associated with increased embryonic mortality.

The relationship between feeding levels and plasma progesterone concentrations was also dealt with by Prime *et al.* (1988), who showed evidence of lower progesterone levels on high food intakes in early pregnancy. It is generally accepted that variations in feed intake towards the end of the pregnancy period are more likely to influence birthweights of piglets than the litter size.

7.5.3. Effect of vitamins

There have been several reports implicating the role of specific dietary ingredients, such as vitamins, in embryonic survival in pigs. Evidence of the importance of folic acid in the diet of multiparous sows in the period between weaning and mating was recorded by Matte *et al.* (1984). Observations recorded in gilts by Bazer and Zavy (1988) indicated that riboflavin may play a specific role in the early development of the embryo; a daily supplement of 100mg to a maize–soybean diet over a short period (4–8 days after oestrus) led to an increase in riboflavin concentrations in uterine flushings and a significant improvement in embryo survival. Vitamin A and β-carotene has also been shown to increase litter size in the multiparous sow. Attempts have been made in the USA by Britt *et al.* (1992) and in Australia by Robertson *et al.* (1995) to understand the mechanisms involved. The American workers concluded that vitamin A might increase litter size by altering development of oocytes or preattachment embryos so that there was greater synchrony between uterine events and embryonic development.

7.6. Fraternity Size and Litter Size

During the 1970s, some intriguing information was reported on the way in which the early postnatal environment of the gilt might influence the size of the litter which it produced in later life. Rearing gilts in small fraternities (six or so pigs in a group) resulted in a phenotypic increase of about one piglet at first parity (Nelson and Robinson, 1976; Rutledge, 1980a,b). It was not obvious what mechanism might be operating in the early weeks of a piglet's life to produce a change in litter size in the animal when it grew up and starting reproducing many months later.

There is evidence to suggest that the pig may differ from other farm mammals in its pattern of oogenesis (Fig. 7.4). Oogonia and oocytes are formed during the first half of fetal life in the cow and ewe; in the sow, evolution is apparently slower, with oogenesis only being completed in the days after birth. Some reports indicated oogenesis lasting until at least 35 days after birth (Fulka *et al.*, 1972); only about 50% of the germ-cell population may survive the transformation of oogonia to oocytes in the early weeks of the piglet's life. As oogonia disappear from the ovaries, the oocytes formed during prenatal life and in the early weeks after birth are the only ones available to the pig throughout its reproductive life.

Rutledge (1980a) suggested that a small fraternity size might provide conditions within the ovary whereby a larger proportion of primordial follicles survive and that this might be a factor influencing the ovulation rate of the

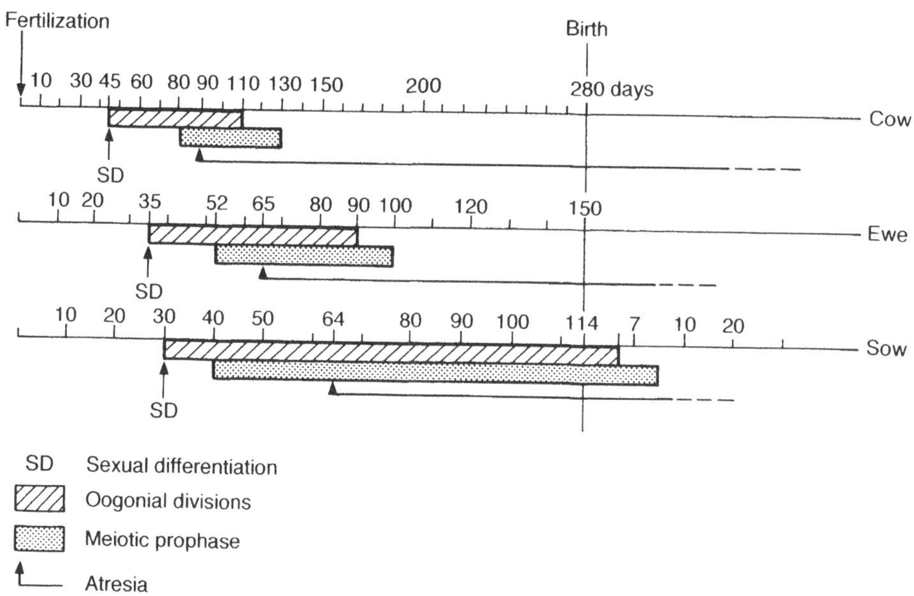

Fig. 7.4. Species differences in the pattern of oogenesis as shown by the cow, ewe and sow (after Thibault and Levasseur, 1974).

animal in later life. From a biological viewpoint, it is of obvious interest to understand some of the ways in which a species may regulate its populations in nature.

7.7. References

Aherne, F.X. and Kirkwood, R.N. (1985) Nutrition and sow prolificacy. *Journal of Reproduction and Fertility* Suppl. 33, 169–183.

Anderson, L.L. (1978) Growth, protein content and distribution of early pig embryos. *Anatomical Record* 190, 143–154.

Anderson, L.L. and Melampy, R.M. (1972) Factors affecting ovulation rate in the pig. In: Cole, D.J.A. (ed.) *Pig Production*. Butterworths, London, pp. 329–368.

Ashworth, C.J. (1990) Effect of pre-mating nutritional status and post-mating progesterone supplementation on early pregnancy in Large White gilts. *Animal Production* 50, 560.

Ashworth, C.J. (1991) Embryo development. *Pig News and Information* 12, 551–554.

Ashworth, C.J. (1991) Effect of pre-mating nutritional status and post-mating progesterone supplementation on embryo survival and conceptus growth in gilts. *Animal Reproduction Science* 26, 311–321.

Baidoo, S.K. and Aherne, F.X. (1988) Sow weight and backfat loss in lactation; effects on the occurrence and the endocrinology of the post weaning estrus. *Proceedings of the 11th International Congress on Animal Reproduction and AI* (Dublin), 2, Paper no. 7.

Baker, D.J., Peters, A.R. and Gettinby, G. (1994) Effect of Altrenogest on litter size in primiparous sows. *Proceedings of the British Society for Animal Production* (Winter Meeting), Paper no. 171.

Bazer, F.W. and Zavy, M.T. (1988) Supplemental riboflavin and reproductive performance of gilts. *Journal of Animal Science* 66 (Suppl. 1), 324.

Bazer, F.W., Martinat-Botte, F. and Terqui, M. (1988) Conceptus development in Large White and prolific Chinese Meishan pigs. *Journal of Reproduction and Fertility* 84, 37–42.

Beltranena, E., Foxcroft, G.R., Aherne, F.X. and Kirkwood, R.N. (1991) Endocrinology of nutritional flushing in gilts. *Canadian Journal of Animal Science* 71, 1063–1071.

Britt, J.H., Lee Whaley, S. and Hedgpeth, V.S. (1992) Improvement of embryo survival by injection of vitamin A in gilts fed normal or high-energy diets before and after mating. *Journal of Animal Science* 70 (Suppl. 1), 271.

Brooks, P.H. (1976) Nutritional problems and sow reproduction. *Proceedings of Meeting and Joint Refresher Course on Pig Production* (Aberdeen), pp. 17–27.

Brooks, P.H. and Cole, D.J.A. (1972) Studies in sow reproduction. 1. The effect of nutrition between weaning and remating on the reproductive performance of primiparous sows. *Animal Production* 15, 259–264.

Brooks, P.H. and Cooper, K.H. (1972) Short-term nutrition and litter-size. In: Cole, D.J.A. (ed.) *Pig Production*. Butterworths, London, pp. 385–398.

Brown, R.W., Hungerford, J.W., Greenwood, P.E., Bloor, R.J., Evans, D.F., Tsonis, C.G. and Forage, R.G. (1990) Immunization against recombinant bovine inhibin α subunit causes increased ovulation rates in gilts. *Journal of Reproduction and Fertility* 90, 199–205.

Buttle, H.L. (1991) Some aspects of endocrinology of reproduction and lactation in pigs. *Pig News and Information* 12, 547–549.

Cassar, G., King, W.A. and King, G.J. (1994) Influence of sex on early growth of pig conceptuses. *Journal of Reproduction and Fertility* 101, 317–320.

Christensen, C., Francis, S., Kowalski, J., Redmond, M. and Laarveld, B. (1994) Increasing gilt reproductive efficiency by immunizing against follistatin. *Biology of Reproduction* 50 (Suppl. 1), 163.

Clark, L.K. (1986) Factors influencing live litter size. In: Morrow, D.A. (ed.) *Current Therapy in Theriogenology*. W.B. Saunders, Philadelphia, pp. 928–930.

Clark, L.K. and Leman, A.D. (1986) Factors that influence litter size in pigs: Part I. *Pig News and Information* 7(3), 303–310.

Claus, R. and Weiler, U. (1985) Influence of light and photoperiodicity on pig prolificacy. *Journal of Reproduction and Fertility* Suppl. 33, 185–197.

Clowes, E.J., Aherne, F.X. and Foxcroft, G.R. (1994) Effect of delayed breeding on the endocrinology and fecundity of sows. *Journal of Animal Science* 72, 283–291.

Cox, N.M., Stuart, M.J., Althen, T.G., Bennett, W.A. and Miller, H.W. (1987) Enhancement of ovulation rate in gilts by increasing dietary energy and administering insulin during follicular growth. *Journal of Animal Science* 64, 507–516.

Crombie, P.R. (1970) Ultra-structure of the foetal–maternal attachment in the pig. *Journal of Physiology (London)*, 210, 101–102.

Day, B.N., Romack, F.E. and Lasley, J.F. (1963) Influence of progesterone–estrogen implants on early embryonic mortality in swine. *Journal of Animal Science* 22, 637–639.

deSa, W.F., Pleumsamran, P., Morcom, C.B. and Dukelow, W.R. (1981) Exogenous steroid effects on litter size and early embryonic survival in swine. *Theriogenology* 15, 245–255.

Dial, G.D. and Bevier, G.W. (1988) Noninfectious causes of reproductive failure in the female pig. In: Laing, J.A., Brinley Morgan, W.J. and Wagner, W.C. (eds) *Fertility and Infertility in Veterinary Practice*, 4th edn. Baillière Tindall, London, pp. 120–139.

Dobao, M.T., Rodriganez, J. and Silio, L. (1983) Seasonal influences on fecundity and litter performance characteristics in Iberian pigs. *Livestock Production Science* 10, 601–610.

Dziuk, P.J. (1992) Embryonic development and fetal growth. *Animal Reproduction Science* 28, 299–308.

Falceto, M.V., Espinosa, E., Gil, L., Josa, A., Cruz, J.L. and Garcia San Clemente, M. (1993a) Environmental effects (hours of daylight, humidity and temperature) on the seasonal variation in ovarian activity). *Proceedings of the 5th International Symposium Animal Reproduction* (Portugal), 2, 425–431.

Falceto, M.V., Espinosa, E., Gil, L., Joas, A., Cruz, J.I., Garcia Martinez, M.P. and Munoz, I. (1993b) Seasonal changes in ovulation rate in sows. *Proceedings of the 5th International Symposium Animal Reproduction* (Portugal), 2, 415–418.

Flowers, B., Martin, M.J., Cantley, T.C. and Day, B.N. (1989) The effect of pregnant mare serum gonadotrophin on follicle stimulating hormone and oestradiol secretion in the prepubertal gilt. *Animal Reproduction Science* 21, 93–100.

French, L.R., Rutledge, J.J. and First, N.L. (1979) Effect of age and parity and litter size in pigs. *Journal of Reproduction and Fertility* 57, 59–60.

Fu, S.L., Dial, G.D., Keister, D.M. and Butler, W.R. (1990) Increased ovulation rate in gilts after oral administration of epostane. *Journal of Reproduction and Fertility* 90, 297–304.

Fulka, J., Kopecny, V. and Trebichavsky, J. (1972) Studies on oogenesis in the early post-natal pig ovary. *Biology of Reproduction* 6, 46–50.

Galvin, J.M., Wilmut, I., Day, B.N., Ritchie, M., Thomson, M. and Haley, C.S. (1993) Reproductive performance in relation to uterine and embryonic traits during early gestation in Meishan, Large White and crossbred sows. *Journal of Reproduction and Fertility* 98, 377–384.

Hammond, J. (1921) Further observations on the factors controlling fertility and foetal atrophy. *Journal of Agricultural Science, Cambridge* 11, 337–366.

Heidler, W., Nowak, P., Knuppel, K., Tschirner, R. and Huhn, U. (1979) Experimental application of steroids to sows in early gravidity to increase litter size. *Archiv fur Experimentelle Veterinarmedizin* (Leipzig) 33, 639–644.

Hooper, P.N., Walters, J.R. and Gray, J. (1982) Identification of the optimum time for service – a review of results with the Walsmetta. *Proceedings of the International Pig Veterinary Society* 7, 180.

Hutchinson, L.A., Findlay, J.K., de Vos, F.L. and Robertson, D.M. (1987) Effects of bovine inhibin transforming growth factor beta and bovine activin-A on granulosa cell differentiation. *Biochemical and Biophysical Research Communications* 146, 1405–1412.

Johnson, R.K., Omtvedt, I.T. and Walters, L.E. (1978) Comparison of productivity and performance for two-breed and three-breed crosses in swine. *Journal of Animal Science* 46, 69–74.

Johnson, R.K., Zimmerman, D.R., Lamberson, W.R. and Sasaki, S. (1985) Influencing prolificacy of sows by selection for physiological factors. *Journal of Reproduction and Fertility* Suppl. 33, 139–149.

King, R.H. and Dunkin, A.C. (1986) The effect of nutrition on the reproductive performance of first litter sows. 3. The response to graded increases in food intake during lactation. *Animal Production* 42, 119–125.

King, B.F., Britt, J.H., Esbenshade, K.L., Flowers, W.L., Sesti, L.A.C. and Ireland, J.J. (1993) Ovulatory and endocrine responses after active immunization of gilts against a synthetic fragment of bovine inhibin. *Journal of Animal Science* 71, 975–982.

Kirkwood, R.N. and Thacker, P.A. (1988) Nutritional factors affecting embryo survival in pigs (results and speculation). *Pig News and Information* 9, 15–21.

Knox, R.V. and Zimmerman, D.R. (1995) Development and evaluation of a treatment regimen for superovulation and embryo survival in pigs. *Theriogenology* 43, 251.

Lodge, G.A. and Hardy, B. (1968) The influence of nutrition during oestrus on ovulation rate in the sow. *Journal of Reproduction and Fertility* 15, 329–332.

Love, R.J., Peacock, A.J. and Evans, G. (1990) Premature farrowings caused by feeding cottonseed meal. *Australian Veterinary Journal* 67, 223–226.

Matte, J.J., Girard, C.L. and Brisson, G.J. (1984) Folic acid and reproductive performance of sows. *Journal of Animal Science*, 59, 1020–1025.

Mattioli, M., Taschini, S., Galeati, G. and Seren, E. (1983) Possibility of assessing the incidence of fetal mortality in sows by early pregnancy diagnosis. *Selezione Veterinaria* 24, 1333–1337.

Mauget, R. (1982) Seasonality of reproduction in the wild boar. In: Cole, D.J.A. and Foxcroft, G.R. (eds) *Control of Pig Reproduction*. Butterworths, London, pp. 509–526.

McGovern, P.T., Morcom, C.B. and Dukelow, W.R. (1981) Chorionic surface area in conceptuses from sows treated with progesterone and oestrogen during early pregnancy. *Journal of Reproduction and Fertility* 61, 439–442.

Mirando, M.A., Harney, J.P., Beers, S., Pontzer, C.H., Torres, B.A., Johnson, H.M. and Bazer, F.W. (1990) Onset of secretion of proteins with antiviral activity by pig conceptuses. *Journal of Reproduction and Fertility* 88, 197–203.

Morgan, G.L. (1987) The effects of altered uterine secretion and utero-blastocyst synchrony on blastocyst development of the pig. *Dissertation Abstracts International B (Sciences and Engineering)* 48(1), 62.

Morcom, C.B., Wildt, D.E. and Dukelow, W.R. (1976) Progesterone : estrone injections during gestation in swine for increasing litter-size. *Proceedings of the International Veterinary Pig Congress*, D25.

Nelson, R.E. and Robinson, O.W. (1976) Effects of post-natal maternal environment on reproduction in pigs. *Journal of Animal Science* 43, 71–77.

Noguera, J.L., Felgines, C. and Legault, C. (1983) Evolution de 1972 a 1981 de composantes de la productivite numerique des truies dans 325 troupeaux francais. *Journees Recherche Porcine en France* 15, 37–51.

O'Grady, J.F., Lynch, P.B. and Kearney, P.A. (1983) Mating management of sows. *Irish Journal of Agricultural Research* 22, 11–19.

Pomeroy, R.W. (1960) Infertility and neonatal mortality in the sow. III. Neontal mortality and foetal development. *Journal of Agricultural Science, Cambridge* 54, 31–56.

Pope, W.F. (1992) Embryogenesis recapitulates oogenesis in swine. *Proceedings of the Society for Experimental Biology and Medicine* 199, 273–281.

Pope, W.F., Lawyer, M.S. and First, N.L. (1987) The effect of exogenous estradiol on litter size in a typical swine herd. *Theriogenology* 28, 9–14.

Prime, G.R., Varley, M.A. and Symonds, H.W. (1988) The effect of food intake during lactation and early pregnancy on plasma progesterone concentrations and prolificacy in multiparous sows. *Animal Production* 46, 499.

Rahnefeld, G.W. and Swierstra, E. (1970) Influence of the sire on litter size in swine. *Canadian Journal of Animal Science* 50, 671–675.

Rampacek, G.R., Robinson, O.W. and Ulberg, L.C. (1975) Uterine capacity and progestin levels in superinducted gilts. *Journal of Animal Science* 41, 564–567.

Reed, H.C.B. (1982) Artificial insemination. In: Cole, D.J.A. and Foxcroft, G.R. (eds) *Control of Pig Reproduction*. Butterworths, London, pp. 65–90.

Rhodes, M.T., Davis, D.L. and Stevenson, J.S. (1991) Flushing and altrenogest affect litter traits in gilts. *Journal of Animal Science* 69, 34–40.

Robertson, J.A., Fairclough, R.J. and Campbell, R.G. (1995) Plasma concentration of retinol after intramuscular injection of 500,000 IU retynl proprionate. *Biology of Reproduction* 52 (Suppl. 1), 72.

Rutledge, J.J. (1980a) Fraternity size and swine reproduction. I. Effect on fecundity of gilts. *Journal of Animal Science* 51, 868–870.

Rutledge, J.J. (1980b) Fraternity size and swine reproduction. II. Genetical consequences. *Journal of Animal Science* 51, 871–874.

Sheridan, P.J., Austin, F.H., O'Connor, P.J. and Roche, J.F. (1986) Effect of exogenous progesterone and oestrone on litter size in swine. *Veterinary Record* 119, 322–324.

Short, T.H., Eckardt, G.R., Sasaki, S., Rose, M., Vincent, A., McLaren, D.G., Mileham, A. and Plastow, G.S. (1995) Marker assisted selection for litter size in pigs. *Journal of Animal Science* 73 (Suppl. 1), 109.

Simmen, R.C.M., Simmen, F.A., Hofig, A., Farmer, S.J. and Bazer, F.W. (1990) Hormonal regulation of insulin-like growth factor gene expression in pig uterus. *Endocrinology (Philadelphia)* 127, 2166–2174.

Symonds, H.W. and Prime, G. (1989) The influence of volume of food intake by gilts on blood flow in the portal vein and clearance of progesterone from plasma. *Animal Production* 48, 620–621.

Thibault, C. and Levasseur, M.-C. (1974) Reproductive life cycle. In: Hafez, E.S.E. (ed.) *Reproduction in Farm Animals*, 3rd edn. Lea and Febiger, Philadelphia, pp. 82–100.

Tilton, J.E. and Cole, D.J.A. (1982) Effect of triple versus double mating on sow productivity. *Animal Production* 34, 179–182.

Tonetta, S.A. and diZerega, G.S. (1990) Local regulatory factors controlling folliculogeneis in pigs. *Journal of Reproduction and Fertility* Suppl. 40, 151–161.

Toplis, G., Ginesi, M.F.J. and Wrathall, A.E. (1983) The influence of high food levels in early pregnancy on embryo survival in multiparous sows. *Animal Production* 37, 45–58.

Varley, M.A. (1982) The time of weaning and its effects on reproductive function. In: Cole, D.J.A. and Foxcroft, G.R. (eds) *Control of Pig Reproduction*. Butterworths, London, pp. 459–478.

Varley, M.A. and Cole, D.J.A. (1976) Studies in sow reproduction. 5. The effect of lactation length of the sow on the subsequent embryonic development. *Animal Production* 22, 79–85.

Webb, A.J. (1995) Future challenges in pig genetics. *Animal Breeding Abstracts* 63, 731–736.

Wildt, D.E., Culvert, A.A., Morcom, C.G. and Dukelow, W.R. (1976) Effect of administration of progesterone and oestrogen on litter size in pigs. *Journal of Reproduction and Fertility* 48, 209–211.

Wu, M.C., Hentzel, M.D. and Dziuk, P.J. (1987) Relationships between uterine length and number of fetuses and prenatal mortality in pigs. *Journal of Animal Science* 65, 767–770.

Wu, M.C., Chen, Z.Y., Jarrell, V.L. and Dziuk, P.J. (1989) Effect of initial length of uterus per embryo on fetal survival and development in the pig. *Journal of Animal Science* 67, 1767–1772.

Embryo Transfer and Associated Techniques in Pigs

8

8.1. Introduction

Effective surgical procedures for the collection and transfer of pig embryos were established by the mid-1960s (Hancock and Hovell, 1962; Dziuk *et al.*, 1964); since that time, the embryo transfer (ET) technique has been used by many workers in research (reviews by Dziuk and Day, 1977; Polge, 1982; Sreenan, 1988). Although the first recorded successful transfer in pigs was reported by Kvasnicki (1951) in the former Soviet Union, much of the research and development work in this species over the years has taken place in Europe, North America and Australasia.

8.1.1. Disease control

Polge (1980) noted that the most important practical application of the ET technique in pigs was probably in relation to disease control; the technique permitted greater flexibility in the introduction of new genetic material into 'closed' herds. As shown by Curnock *et al.* (1976), ET could be used very effectively as an alternative to hysterectomy in getting disease-free piglets into a specific-pathogen-free (SPF) herd. The donor sow was not lost, as with hysterectomy, and the risks of introducing infectious disease were seen to be less with embryo transfer.

It is known that hysterectomized-derived piglets are not automatically free from infection. It is possible for them to carry a variety of pathogens, especially those that can cross the placenta, such as porcine parovirus (PPV), swine fever virus (SFV), Aujeszky disease virus (ADV), leptospira and brucella (Wrathall, 1984). Also pig ET could be useful where there had been a serious breakdown in the health status of a 'closed' herd and the technique was used in 'salvaging' valuable genetic material by transferring embryos into a 'clean' herd located elsewhere.

Table 8.1. Primary reasons for using embryo transfer in pigs in the USA in 1981 (from Martin, 1986).

Group no.	Reason	No. of recipients (%)	No. of herds
1	Establish new herds from herds with pseudorabies	113 (43)	4
2	Make additions to specific-pathogen-free (SPF) herds	67 (26)	13
3	Obtain boars for closed commercial herd	61 (23)	11
4	Obtain more offspring from superior gilts and sows	19 (7)	8
		260	36

In the USA, Martin (1986) also mentioned disease control/prevention as the primary reason for the commmercial application of pig ET in that country in the early 1980s (Table 8.1). In Australia, Cameron *et al.* (1989) reported on their experiences with commercial embryo transfer which was used to introduce new genetic material into five SPF pig herds. In this work, 751 embryos were collected surgically from 37 gilts and sows and 665 embryos were transferred surgically to 39 synchronized recipient sows within 12 h; 31 (80%) of the pigs farrowed, with an average litter size of 8.1.

According to Wrathall (1984), pig ET at that time was used for three main purposes: (i) for research in reproduction and pathology; (ii) as a means of exploiting the superior genetic characteristics of individual sows; and (iii) for the movement of genetic material with a low risk of concurrent transmission of infectious disease. The same author noted that commercial interest in pig ET was growing mainly in the context of disease control; in terms of the international movement of pig embryos, guidelines laid down by the veterinary animal health authorities in various countries should ensure that ET was much safer than any other means of movement of pigs.

8.1.2. International trade

In the context of international trade, the importance and significance of the embryo's zona pellucida (ZP) requires special mention. The membrane has assumed particular significance because many viruses can be readily washed or digested off the ZP by trypsin treatment, thereby removing or minimizing the risk of transmitting disease along with the embryo. Most countries require that imported embryos possess an intact zona pellucida and that it has passed through an approved washing procedure. As far as pigs are concerned, there appear to be species differences in the way that viruses adhere to the ZP; it has proven more difficult to remove viruses from the ZP of pigs than of cattle (Singh, 1987).

8.1.3. Breeding improvement and other advantages

The usual reason for interest in embryo transfer in the farm animals is in order to obtain as many young as possible from the genetically outstanding female; this can apply in pigs, but not in the same way as in cattle and sheep. In normal pig-breeding programmes, the emphasis is generally on the genetic merits of certain strains or lines rather than on individual pigs. The high reproductive rate of the sow also means that it can produce as many young in a 6-month period as a cow or ewe might produce in its entire reproductive lifetime. There are limited circumstances in which the technique may be applied to the individual sow. It may, for instance, be possible for a valuable but aged sow to serve as a donor of several batches of embryos when she has gone beyond the point of carrying a litter to term and rearing them herself.

8.1.4. Difficulties in cryopreservation

The export–import side of embryo transfer in pigs has been held up by the fact that pig embryos could not be frozen in the same way as those of cattle and sheep. However, as discussed in section 8.4 there has been much research activity in developing procedures which may overcome that problem. There have been limited reports of pig embryos being shipped around the world in the fresh state; instances recorded in the literature include the transport of embryos between the USA and Spain (James and Reeser, 1979) and between Canada and the UK (Wrathall *et al.*, 1970). A further example was that of James *et al.* (1980) in which embryos recovered surgically from Chester White and Hampshire pigs in the USA were shipped to the UK and there transferred to Large White recipient gilts; in all, 227 embryos were transferred to 12 recipients, resulting in the birth of seven litters and a total of 58 piglets.

Japanese workers have suggested that embryo transfer technology could be used for the international exchange of miniature pig strains; these small pigs are used as laboratory animals in medical and dental research. Yamakawa *et al.* (1990) transferred embryos from Gottingen miniature pigs to two normal domestic pig recipients and reported the birth of 21 piglets.

8.1.5. Requirements for surgery

The commercial application of embryo transfer in pigs has been restricted by the requirement for tedious and costly surgical interventions in the recovery and transfer of the embryos. In this, the story in pigs is quite different from that in the cow, where effective non-surgical recovery and transfer techniques were rapidly developed. Apart from that, welfare concerns in exposing the pig to such procedures must be taken seriously; all the more reason for concentrating efforts on the development of acceptable non-surgical methods. It is now possible for pig embryos to be produced in the laboratory where previously they could only be recovered by flushing the tubes and uterus of the recently bred sow; the source of oocytes used in this may be the ovaries of pigs after slaughter (see section 8.6).

In Germany, Schlieper and Holtz (1986) investigated the suitability for transfer of two- to four-cell pig embryos flushed from the tract of slaughtered animals. They reported that success rates when transferring embryos collected surgically or from the reproductive tracts of slaughtered pigs did not differ significantly, provided the tracts were removed and flushed immediately after slaughter. Wollenberg *et al.* (1990) also reported on the recovery of four- to eight-cell embryos from the reproductive tracts of gilts slaughtered 78 h after the expected time of ovulation; they similarly stressed the importance of embryo recovery immediately after slaughter.

8.2. Superovulation Techniques

Superovulation may be induced with two main objectives in mind: (1) to provide embryos for subsequent transfer; and (2) to provide oocytes or early embryos for research purposes (e.g. gene transfer studies).

8.2.1. Cyclic animals

Although the sow normally releases several oocytes during oestrus, there is still likely to be a need to induce an increase in the ovulation rate for embryo transfer purposes. Increased ovulation rates have been obtained in the postpubertal gilt and sow by administering a single dose of PMSG at the beginning of the follicular phase of the oestrous cycle (optimum stage appears to be days 15 and 16). There appeared to be a positive linear relationship between dose of PMSG and number of ovulations; Hunter (1964) calculated a regression coefficient of 1.89 + 0.5 corpora lutea per 100 IU PMSG. The gonadotrophin treatment usually reduced the length of the oestrous cycle by about 1 day and increased the duration of the heat period; normal fertilization rates could be expected in the superovulated pigs bred by the usual procedures (Phillipo, 1968; Guthrie *et al.*, 1974).

In a study by Kim *et al.* (1990) cyclic pigs were superovulated with PMSG (1000 IU) and hCG (500 IU) on days 15–16 of the cycle and embryos were recovered some 4–6 days after the hCG injection. For collections on days 4, 5 and 6, the average number of embryos collected was 13.1, 17.1 and 12.7, respectively. The same authors provided information on responses after the repeated superovulation of donor animals. In China, Chen and Wu (1991) described the procedures used in their laboratory, including anaesthesia, laparotomy and embryo collection; they recorded a recovery rate of >80% in 90% of gilts. In Missouri, Galvin *et al.* (1994) induced superovulation in cyclic gilts with 1000 IU PMSG administered on day 16 of the cycle followed by

500 IU hCG after three days; donors were checked daily for oestrus and bred at least twice by the boar or by AI on day 0 and every 24 h while in heat.

8.2.2. Prepubertal gilts

For research purposes and for developing embryo transfer technology in pigs, prepubertal gilts were often regarded as an inexpensive source of embryos. These animals also had the advantage that the superovulation treatment was not complicated by the presence of an oestrous cycle, which meant that gonadotrophin treatment could be applied at any time. Doses of 250–2000 IU PMSG, in combination with hCG injected 48 h later, were used in studies by Baker and Coggins (Table 8.2); hCG was given in an effort to improve the precision of ovulation.

Provided prepubertal gilts were reasonably well grown (>80 kg), Baker (1979) showed that treatment with a dose of 1000 IU PMSG, followed 72 h later with 500 IU hCG, and then insemination at 24 h beyond that, could give satisfactory results; about 90% of ovulations were accounted for in recoveries and 60% of oocytes were fertilized. With an average ovulation rate of 25, the author recovered 22 oocytes, of which 13 were fertilized. Work elsewhere with prepubertal gilts reported evidence of a marked seasonal variation in response to PMSG; ovulatory responses proved to be twice as high in spring as at other times of the year (Day, 1979).

In Germany, Springmann *et al.* (1988) studied various dose levels of PMSG (750–1500 IU) with hCG (750 IU) given 3 days later. They analysed the influence of several factors on the superovulatory response and on embryo quality: (i) bodyweight of donor; (ii) the PMSG dosage; (iii) the month of oocyte collection; (iv) the quality of the induced oestrus; (v) the boar used. With an increase from 16 to 24 IU PMSG kg^{-1} bodyweight, the average number of transferable quality embryos increased from 10.9 to 24.6; some differences were evident on a seasonal basis according to the boar used.

In Taiwan, Kang and Wu (1991) studied various superovulation treatments in both prepubertal and postpubertal gilts. The prepubertal animals were given 1000 IU PMSG subcutaneously followed 72 h later by hCG (500 IU) given intramuscularly. Cyclic gilts received PG after day 12, followed

Table 8.2. Superovulatory responses in prepubertal gilts in relation to the dose levels of PMSG used (source: Baker and Coggins, 1968).

PMSG	hCG	Mean no. of ovulations
250	500	7.2
500	500	12.5
1000	500	19.6
2000	500	45.8

48 and 120 h later by PMSG and hCG, respectively. It was concluded that, regardless of age, gilts could be superovulated with 1000 IU PMSG and 500 IU hCG. In Germany, Reichenbach *et al.* (1993) treated prepubertal gilts with 1500 IU PMSG, followed 72 h later by 750 IU hCG; donors were bred naturally or inseminated at 24 and 36 h after hCG administration.

8.2.3. Superovulation after preliminary altrenogest treatment

A number of authors have reported using gonadotrophins and $PGF_{2\alpha}$ in combination with a preliminary treatment with altrenogest (Regumate). In Taiwan, Kang and Wu (1991) administered altrenogest (20 mg daily) for 18 days to postpubertal gilts, with PMSG given on day 19 and hCG after a further 72 h. In the Czech Republic, Kubicova *et al.* (1993) used altrenogest over a 19-day period and then followed with different PMSG and hCG dosages; it was concluded that treatment with 2000 IU PMSG in combination with 600 IU hCG (given after 78 h) gave the best results.

In Canada, Bolamba and Sirard (1995) compared the efficacy of superovulation in cyclic gilts synchronized with altrenogest and prepubertal gilts primed with exogenous gonadotrophin (PG600) followed by synchronization with prostaglandin before induction of superovulation. On the basis of embryo culture studies, it appeared that the prepubertal gilts were the superior embryo donors.

In pigs used to supply oocytes for research, Pursel and Wall (1995) regulated oestrus in the animals by altrenogest. Donors were injected with PMSG (1500–2000 IU) at 24 h and hCG (500 IU) at 103 h after the last feeding of altrenogest.

8.2.4. Effect of repeated superovulation

A number of reports have dealt with repeated embryo recoveries in pigs. In Japan, Hayashi *et al.* (1994) dealt with the effects of recovering embryos from naturally ovulated donor sows in 4 to 16 occasions per sow. They concluded that recovery at intervals of 3 weeks did not adversely affect the recovery process. Kim *et al.* (1990) made reference to repeated recoveries in their studies in Korea.

8.2.5. Breeding the donor pig

For optimal fertilization rates, it is usually recommended that donor pigs should be mated or inseminated every 12 h throughout oestrus (Martin, 1986). If hCG has been employed to control the time of ovulation, the important times for insemination with fresh and frozen semen would be at 24 and 36 h, respectively, after the ovulating hormone. The normal considerations for breeding by AI would be applied to the donor pig; the volume of the inseminate should be 50–100 ml and it should contain at least 4000 to 5000 million live sperm.

8.3. Embryo Recovery Procedures

8.3.1. Surgical recoveries

Methods employed in the surgical collection of pig embryos are much the same today as when first developed in the 1960s. Embryo recovery is generally carried out by way of a mid-ventral laparotomy under general anaesthesia induced by a barbiturate, and usually maintained with halothane using a closed circuit system. The reproductive tract is exteriorized with manipulation of the oviducts and uterine horns kept to a minimum; aseptic precautions are taken at all stages of the recovery process.

Oviductal oocytes can be collected after flushing the tube from the uterotubal junction (UTJ) through the cannulated infundibular end of the oviduct with 20–30 ml of warm flushing fluid (Day, 1979). The flushing medium has to be introduced into the oviduct and not the uterine horn, since, in the pig, the structure of the UTJ does not permit backflushing into the oviduct from the uterus. In species with intrauterine ejaculation, such as the pig, it is believed that the UTJ serves as a barrier to the gross passage of seminal plasma into the Fallopian tubes (Hunter, 1988). In this regard, the pig's UTJ could be regarded as analogous to the cervix in preventing access to the uterus in species with intravaginal ejaculation, such as cattle and sheep. The movement of pig embryos through the oviduct is more rapid than in cattle and sheep and they normally enter the uterus at the four-cell stage about 48 h after ovulation (Oxenreider and Day, 1965).

8.3.2. Collecting embryos for transfer

Four-cell embryos are frequently required for transfer and these can be recovered 4 days after the onset of heat in most sows (Martin, 1986); at this time, however, the embryos may be located in either the oviduct or the tip of the uterine horn. The usual method employed to recover such embryos involved flushing fluid through the oviduct from the ovarian end into the uterine horn and trapping it in a segment of the uterus; a cannula was then inserted through the uterine wall to recover the fluid and embryos (Hancock and Hovell, 1962; Dziuk *et al.*, 1964). Modifications of the procedure were described by Day (1979), including one in which a silicone rubber catheter was inserted into the oviduct 5–10 mm above the UTJ and 1.5–2.0 cm of the tube was threaded into the uterine horn; flushing fluid was then injected into the uterus and collected back through the catheter.

Using techniques such as those described, within 6 days of ovulation, it proved possible to achieve embryo recovery rates of the order of 80–90% (Polge, 1977a,b; Day, 1979); repeated collections (five times in 90 days) were reported by some workers (James and Reeser, 1979). However, the length and nature of the sow's cervix and uterine horns appeared to rule out the possibility of using non-surgical recovery procedures.

8.3.3. Transcervical collection of embryos

A method for the transcervical recovery of pig embryos, reported by Hazeleger *et al.* (1989) in the Netherlands, was based on a procedure to shorten the uterine horns. They describe the transcervical flushing of four sows with resectioned uteri and record that an average of three embryos was recovered.

8.3.4. Evaluation of pig embryos

Hancock (1961) estimated that the first cleavage division in pigs occurs about 20 h after the initial fertilization events and the second embryonic cycle takes about 12 h.

Fertilization and early embryonic development in the pig has been reported in some detail (Hunter, 1974). Cleavage from the single cell to the four-cell stage occurs in the oviduct during the first 36 h after fertilization; embryos enter the uterus at the four- to six-cell stage. On the basis of observations on several thousand pig embryos at Cambridge, Polge (1977a,b) recorded that development could be arrested at the four-cell stage for up to 48 h; four-cell embryos were not uncommon when recovery was made as late as 70–80 h after ovulation. When cleavage resumed, division occurred about every 12 h, but by no means all four-cell embryos continued development together. For that reason, it was not unusual to recover four-cell embryos in the presence of those consisting of 8 to 16 cells.

Kronnie and Boer (1993) recorded that the four-cell stage in their studies lasted about 20 h. In agreement with Jarrell *et al.* (1991), they concluded that activation of the embryonic genome (transition from maternal to zygotic control of development) occurred during the four-cell stage in the pig. Schoenbeck *et al.* (1993) indicated that the maternal genome controls development until about 12–16 h into the four-cell stage; the embryonic genome subsequently begins to control development at the G2 phase of the cell cycle. According to Polge (1977a,b), compaction of the pig embryos can occur as early as the eight-cell stage, with the blastomeres becoming fused together and making it impossible to count accurately the number of cells present, other than by a process of fixing and staining.

Variation in embryo sizes

Considerable variation in size among pig embryos may be evident after the hatching of blastocysts from the zona pellucida, which takes place between the sixth and seventh day after ovulation; it is uncertain whether such differences influence the survival of the embryos subsequently. It seems unlikely that the duration of ovulation in sows plays a part in this. Soede and Kemp (1993), for example, found the duration varied between 1 and 4 h in their studies. The same authors did find that within-litter embryo diversity was not related to ovulation duration but was negatively correlated with accessory sperm count; the indications were that duration of fertilization could affect embryo diversity.

There have been suggestions that variation in the rate of early embryo development within a litter may be a cause of the high incidence of embryo mortality in European breeds of pig (Pope *et al.*, 1986, 1990). There are also studies reporting an increased uniformity of blastocyst development in the Chinese Meishan breed and the suggestion that this may be a factor contributing to the breed's exceptional prolificacy (Bazer *et al.*, 1988).

In the Meishan context, it should be noted that Youngs *et al.* (1991, 1993) in Iowa found that Meishan embryos were significantly slower in reaching the compact morula and blastocyst stage than Large White embryos.

In terms of the most appropriate cleavage stage for collecting and transferring pig embryos, authors differ. Martin (1986) found that most embryo transfer specialists prefer to collect embryos from donors when they are at the four- to eight-cell stage.

8.3.5. The fragmentation phenomenon

One feature of pig embryos which is seldom encountered in cattle and sheep is that of fragmentation. This fragmentation occurs in unfertilized oocytes, rarely during the first 40 h after ovulation when they are in the Fallopian tubes, but commonly after this time when they are in the uterus (Dziuk, 1960). Fragmentation can take the form of a very even cleavage to two cells, but is more usually an uneven fragmentation of the cytoplasm.

8.4. Storage and Cryopreservation of Pig Embryos

Storage of pig embryos is essential, whether in simply holding them for a short period before transferring to a recipient or in holding them in a viable state while they are transported around the world. The international transportation of pig embryos, held at 36.5°C in 0.25 ml plastic straws, was recorded by Niemann (1989); the embryos were morulae and held in culture medium enriched with serum. After a 30–34 h journey from Germany in a Thermos container, most embryos were viable and many gave rise to piglets on transfer in Argentina. In a second trial, pig embryos were transported by train and car from Germany to the former Yugoslavia; after 19–23 h in the Thermos flask, 95.3% of the embryos were morphologically intact and on transfer many gave rise to normal piglets. However, such transportation of embryos in the fresh state is very much the exception; most international trade in farm animal embryos is based on shipping them in the frozen state. For the pig, unfortunately, the freezing of embryos has posed extremely difficult problems.

8.4.1. Short- and longer-term storage of embryos

In many of the studies dealing with embryo transfer in the pig, it has been a matter of transferring the embryos into appropriate recipients with a minimum of time spent in storage. Polge (1977a,b) described work at Cambridge on the short-term storage of pig embryos. Storage temperature was held to be critical (no survival at storage below +15°C) but it was possible to store four-cell embryos at 20°C for 24 h in suitable media and for them to survive reasonably well after transfer. Culture media in the Cambridge work included Brinster's medium, Medium 199, Ham's F-10 and Dulbecco's phosphate-buffered saline (PBS) enriched with pyruvate and lactate.

8.4.2. Culture of early cell stages

The early workers found that the proportion of two-cell pig embryos that continued development *in vitro* was low, regardless of the culture system employed (Pope and Day, 1977; Wright, 1977; Allen and Wright, 1984). Early development of the embryo was characterized by a diapause at the four-cell stage, which is now recognized as the developmental block (Schoenbeck *et al.*, 1993). Limited success in culturing one- and two-cell embryos through the block was reported, using modified Whitten's culture medium (Menino and Wright, 1982).

Pig embryos at later stages of development (beyond the four-cell stage) could be cultured to the blastocyst *in vitro* but this was often associated with reduced embryo viability. Davis and Day (1978), for example, cultured pig embryos in a modified Krebs Ringer bicarbonate medium (modified KRB) containing combinations of glucose, lactate, pyruvate and bovine serum albumin. Culture for longer than 24 h markedly decreased embryo survival after transfer. Much reduced viability of cultured pig embryos was recorded by Polge (1982) who found that culturing *in vitro* for 48 h resulted in embryonic survival rates of only 20–30%. Literature on *in vitro* embryo culture was reviewed by Davis (1985) and more recently by Reed *et al.* (1992).

8.4.3. Pigs and rabbits

In contrast to what occurred *in vitro* with early pig embryos, their culture in the rabbit or sheep oviduct did permit normal embryonic development (Polge, 1982) when these were later transferred to recipient gilts. It had been shown in pigs (Polge *et al.*, 1972), as had earlier been demonstrated in sheep (Averill *et al.*, 1955) and cattle (Sreenan and Scanlon, 1968), that pig embryos would continue developing when transferred to the ligated oviducts of oestrous and pseudopregnant rabbits; in pigs, a high embryo survival rate (90%) was achieved when embryos incubated in the rabbit for 2 days were re-transferred to recipients.

8.4.4. Pigs and mice

Studies in the early 1960s had shown that mouse oviducts maintained in organ culture could be used to bypass a developmental block in mouse embryos (Biggers *et al.*, 1962; Gwatkin and Biggers, 1963). Subsequently, Krisher *et al.* (1989) reported the ability of the mouse oviduct to support development to the blastocyst stage of early-stage embryos from a number of species, including the pig. Further studies in which most one-cell pig embryos developed to the morula or blastocyst stage, led them to conclude that the isolated mouse oviduct culture system apparently mimics the conditions to which the pig embryo is exposed *in vitro.*

8.4.5. Co-culture with oviductal cells

In the light of reports on the successful culture of sheep and cattle embryos through the developmental block stage by the use of oviductal cells, White *et al.* (1989) in Louisiana evaluated the co-culture of early pig embryos with pig oviductal cells (POEC), pig fetal endometrial fibroblasts (PEF) and a combination of these two cell types. The results indicated that POEC provided factors necessary for embryos to cleave through the four-cell block (only 11% and 7% of embryos cultured in POEC and PEF–POEC failed to develop past the block stage) and subsequently to the hatched blastocyst stage.

Pipkin *et al.* (1991) suggested that pig embryos could be co-cultured with pig oviductal cells through the four-cell block with greater success using BMOC medium supplemented with fetal calf serum (FCS) or bovine serum albumin (BSA) than in other media (Ham's F-10; Ham's F-12; Whitten's medium). In Denmark, however, Smith *et al.* (1993) reported that the oviductal cell co-culture of embryos in Whitten's medium did not support a greater degree of blastocyst development of one- and two-cell pig embryos than when medium alone was employed; the reasons for this were not apparent.

8.4.6. Culture with pig oviductal fluid

Studies reported by Archibong *et al.* (1989) in North Carolina dealt with the development of one- and two-cell pig embryos in culture medium supplemented with pig oviductal fluid. Development of one-cell and two-cell embryos to four-cell, in 48 h and 24 h, respectively, was similar to that recorded by Oxenreider and Day (1965) and Hunter (1974) for *in vivo* development.

8.4.7. Modifications to culture media

At Madison, Hagen *et al.* (1991) reported the successful development of early pig embryos through the block stage in simple media (M-199 and four formulations of modified Tyrode's medium (containing lactate and pyruvate). Their results indicated that pig embryos could be cultured from the one-cell stage to blastocyst in a simple HEPES-buffered medium in air. The authors noted that the ability of pig embryos to develop without supplemental carbon dioxide may be useful in situations in which embryos must be transported for long periods before embryo transfer. In a review of the literature, Reed *et al.* (1992) noted how modifications to simple culture media had demonstrated that glutamine can serve as the sole exogenous energy source for development of the pig zygote to the blastocyst stage *in vitro*. Glucose is apparently not inhibitory to the development of pig embryos *in vitro* in the way that it is to cattle and sheep embryos. The reviewers observed that the addition of taurine and hypotaurine to the medium further increased the degree of development *in vitro*. This information was reported by Petters and Reed (1991) and Illera *et al.* (1992) who had shown that a medium containing glutamine, taurine and hypotaurine would support one-cell pig embryos to the blastocyst stage.

Studies by Beckmann and Day (1993) demonstrated that pig embryos could be successfully cultured past the four-cell block stage in Whitten's medium, which has a relatively low concentration of NaCl. According to Galvin *et al.* (1993), the NaCl concentration of conventional culture media may be responsible for the developmental block observed in pig embryos. They showed that this NaCl-induced block may be avoided by the addition of glutamine. It can be concluded that a number of different methods are now available to culture pig embryos from the one-cell stage *in vitro* as a result of the greater understanding of the requirements of embryos.

8.4.8. Frozen storage of pig embryos

In the freezing of embryos, the methods originally used by Whittingham (1971) and Wilmut (1972) with mice were found to be entirely unsuccessful with pigs. Three other approaches have been taken towards the cryopreservation of the pig embryo (Fig. 8.1). The fact that pig morulae are exceptionally sensitive to cooling below 15°C has been recognized for some years (Polge *et al.*, 1974; Niemann, 1985), but even in relatively recent reports, showing that the lethal effect of chilling on pig morulae occurs abruptly just below 15°C, there has been uncertainty as to the precise nature of the problem (Plante *et al.*, 1993). However, it is becoming increasingly clear that the sensitivity of pig embryos to cooling and freezing is probably the result of their high lipid content.

In Japan, Nagashima *et al.* (1988) concluded that pig embryos at the expanded and hatched blastocyst stage have a higher tolerance to cooling compared with early blastocysts or morulae; blastocysts were shown to

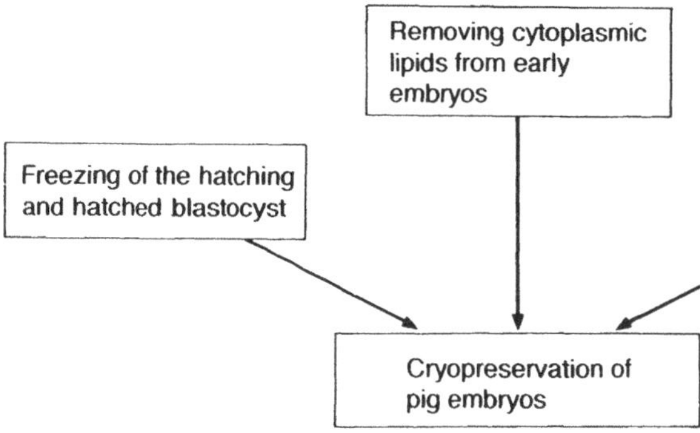

Fig. 8.1. Three approaches to the successful cryopreservation of pig embryos.

maintain viability and to develop into normal fetuses after exposure to a temperature of 11°C, which is below the critical temperature of 15°C at which irreversible damage normally occurred. The freezability of perihatching-stage blastocysts was confirmed by freezing to −20°C, using 1.5 M dimethyl sulphoxide (DMSO) as the cryoprotectant (Nagashima *et al.*, 1989a,b). Hayashi *et al.* (1989) showed that it was possible to store pig embryos at the expanded/hatched blastocyst stage briefly at subzero temperatures with 1.5 M glycerol and to produce five piglets from 11 embryos after thawing. A similar freezing procedure was used with expanded and hatched blastocysts by Feng *et al.* (1991) in China, who recorded the birth of two piglets after the transfer of 42 embryos to a recipient.

The freezing of perihatching-stage pig blastocysts and their subsequent storage in liquid nitrogen was later reported by two Japanese groups (Kameyama *et al.*, 1990; Kashiwazaki *et al.*, 1991); in both instances live piglets were born after transfers. Reviewing *in vitro* culture studies after storage of pig embryos in liquid nitrogen, Nagashima *et al.* (1994) noted survival rates of between 17% and 50% for expanded blastocysts and between 38% and 96% for hatching and hatched blastocysts. Embryos younger than the early blastocyst stage and hatched blastocysts having a diameter larger than 300 μm did not survive freezing, indicating that freezability of the pig blastocyst peaks shortly after hatching and then decreases rapidly.

Changes in the morphology and cell number of the inner cell mass (ICM) of pig embryos as a result of freezing was examined by Iwasaki *et al.* (1994), who found that the ICM at the expanded and hatched blastocyst stages survived even at −196°C and that the degree of ICM damage was lower at the hatched stage than at the expanded stage. Subsequently Iwasaki *et al.* (1995) examined the viability of pig embryos thawed at different temperatures after conventional slow freezing.

8.4.9. Suggested freezing protocol for blastocysts

In the light of research findings, a protocol for the freezing of pig embryos was developed by Nagashima *et al.* (1994). Blastocysts at the perihatching stages must be used; they can be developed from earlier stages using *in vitro* culture with a conventional embryo culture media (e.g. Whittingham's M-16; Whitten's medium; mKRB) to which BSA was added. As a cryoprotectant, 1.5 M glycerol has generally been regarded as effective, although Weber and Youngs (1994) have shown that pig embryos exposed to glycerol or propylene glycol exhibited poorer development than those exposed to ethylene glycol; conventional slow cooling methods can be used (0.3°C min^{-1} from –7 to –35°C). After rapid thawing in a water-bath at 35–37°C, step-wise dilution of the cryoprotectant is employed. An improvement in the survival rate of pig embryos was reported by Nagashima *et al.* (1995a,b) when they cryopreserved hatched blastocysts with a combination of glycerol (1.5 M) and sucrose (0.25%); transfer of 66 embryos to a recipient after thawing resulted in the birth of four piglets.

In talking about the cooling and freezing of pig blastocysts, there have been authors who have examined the effect of heat shock on the tolerance of such embryos to low temperatures. In China, Xu *et al.* (1994) found that heat shock significantly increased the tolerance of pig blastocysts, expanded blastocysts and hatched blastocysts.

8.4.10. Removal of cytoplasmic lipid

Nagashima *et al.* (1994, 1995a,b) have clearly demonstrated that early-cleavage-stage pig embryos survive freezing after removal of cytoplasmic lipid. They also showed that piglets born from the frozen–thawed delipidated embryos were normal in appearance and had birthweights and postnatal growth rates normal for the breed. Such work has shown that the high sensitivity of the pig embryo to cooling is the result of its high lipid content. It has been evident for some years that the lipid droplets abundant in one- to eight-cell stage pig embryos decline markedly at the blastocyst stage (Niimura and Ishida, 1980). There is a further decline in the size of lipid droplets after hatching, which presumably explains the success in freezing embryos at this stage.

From the animal health viewpoint, working with delipidated embryos or hatched embryos still has very obvious disadvantages; the lack of the zona pellucida, the internationally accepted barrier to disease pathogens, currently makes such embryos unacceptable for importation into many countries.

8.4.11. Vitrification

Although studies in the USA (Didion *et al.*, 1990) demonstrated that pig oocytes at the germinal vesicle (GV) stage did not survive cooling to 15°C or below, several authors have reported evidence indicating that vitrification may be a more hopeful approach towards the cryopreservation of pig oocytes (Rubinsky *et al.*, 1991; Arav,1992; Arav *et al.*, 1993; Nagashima *et al.*, 1996). Vitrification is the rapid cooling of a solution without formation of ice crystals. In this way oocytes and embryos are not subjected to the physical damage otherwise associated with crystal formation.

In some of the vitrification studies, survival rate of the oocytes was apparently increased by the addition of antifreeze glycoproteins, derived from Antarctic fish. It is believed that protection is provided to cell membranes by the ability of the antifreeze proteins to interact with or perhaps bind to them. Nagashima *et al.* (1996) have also demonstrated that pig oocytes can be cryopreserved by vitrification after centrifugation to separate the lipid droplets from the cytoplasm; however, with their current protocol, the oocytes appeared to be less viable than delipidated oocytes.

Some preliminary results from work in the US Department of Agriculture (USDA) at Beltsville also indicated that day 6 pig embryos are capable of surviving vitrification (Dobrinsky, 1993; Dobrinsky and Johnson, 1993). The extent and nature of nuclear and cytoskeletal damage that occurred during the vitrification of day 5, 6 and 7 pig embryos was documented by Dobrinsky *et al.*, (1994) who concluded that cytoskeletal analysis may be a useful tool in developing improved media for the cryopreservation of pig embryos. In Australia, Nagashima *et al.* (1996) showed that delipidated two- and four-cell pig embryos could be successfully preserved by vitrification and that they were capable of developing to the blastocyst stage subsequently.

The development of an effective procedure for the freezing of pig oocytes or embryos could have many practical implications and could be expected to result in a significant impact on pig-breeding programmes around the world (see Dobrinsky, 1993).

8.5. Embryo Transfer Methods and Recipient Management

The first report of the birth of litters of normal size after embryo transfer in pigs was that of Hancock and Hovell (1962). However, it is of historical interest to note that successful transfer of embryos between pigs was recorded prior to that time, but the number of piglets born was abnormally low (Kvasnicki, 1951; Pomeroy, 1960). In the USA, embryo transfers are usually carried out on the pig farm rather than in a clinic to reduce the risk of introducing disease. As with surgical recovery procedures, general anaesthesia is induced and maintained throughout the procedure and the reproductive tract is reached through a mid-ventral incision; corpora lutea of the recipient are examined for stage of development before the transfer is made.

8.5.1. Synchronizing donors and recipients

As with cattle and sheep, exposure of the pig embryo to an asynchronous uterus can adversely affect development. Although exact synchrony of the oestrous cycle stage between donor and recipient pigs is desirable, Webel *et al.* (1970) showed that transfers in which the donor was 1 or 2 days earlier or 1 day later than the recipient were apparently just as successful. A later assessment of synchrony requirements (Polge, 1982) showed that there was no reduction in pregnancy rate when donors were 1 day earlier than recipients but transfers to recipients more advanced than donors did result in a decrease. In Missouri, Jarrell *et al.* (1990) assessed the ability of day 4 pig embryos to develop after short-term (24 h) exposure to an asynchronous (day 6) uterine environment; they concluded that such exposure increased the incidence of embryo mortality.

8.5.2. Cell stage at transfer

Although acceptable embryo survival rates were achieved by the collection and transfer of one-cell and two-cell embryos to the oviducts of synchronized recipients (Pope *et al.*, 1972), the use of one-cell embryos did present a difficulty in deciding whether they were fertilized or not. Embryos at the one- and two-cell stage were transferred by surgical procedures to the oviducts of recipient; they were aspirated into a glass pipette (fire-polished tip) in a small volume of transfer medium (e.g. M-199, Dulbecco's PBS) and deposited in the lower ampulla via the fimbriae.

Embryos at the four-cell or later stages of development were transferred to the tip of the uterine horn by means of a pipette, the uterine horn having first been punctured with the blunt end of a suture needle; care was taken to ensure that embryos were deposited into the lumen of the uterus. Another method of transferring embryos to the uterus was by flushing then into the uterus by way of the oviduct (Polge, 1977a); this avoided the need to puncture the uterine wall and the risk of possible bleeding, which might adversely affect embryo survival.

In general, embryo transfer in pigs has been carried out with embryos at the four- to eight-cell stage of development (Day, 1979); normal cleavage is more apparent in the four-cell embryo than at later stages. After the eight-cell stage and the start of compaction, it is often difficult to distinguish individual blastomeres; an incorrect evaluation of normal cleavage or degeneration is more likely than at the four- to eight-cell stage.

Transfers need only be carried out on one side of the uterus; subsequent transuterine migration of the embryos should ensure that they space themselves out in a reasonably uniform manner between the two uterine horns (Polge and Dziuk, 1970). The aim was usually to transfer about 16 embryos per recipient and to expect survival rates of 60–65% (Polge, 1980); such rates compared favourably with those in the normal pig. The fact that embryos at the early cleavage stages (even the two-cell stage) could be successfully transferred to the uterus of the pig was in sharp contrast to what was known to be possible in cattle and sheep.

Embryo age at transfer

On the other hand, cattle and sheep embryos can be successfully transferred as late as day 12 of the oestrous cycle; in pigs, the pregnancy rate was reported as decreasing markedly when transfers were attempted after day 6 (Hunter *et al.*, 1966; Webel *et al.*, 1970). However, a transfer procedure was developed at Cambridge which permitted older embryos to be transferred successfully (Polge, 1982). Transfer of 3- to 7-day-old embryos gave a 75% pregnancy rate but those attempted on days 8 and 9 were less successful. In this procedure, pig embryos were transferred to recipients by a pipette introduced in a small incision at the base of the oviduct and slid down into the uterine horn through the UTJ; the procedure apparently reduced the liklihood of endometrial trauma and haemorrhage which could otherwise prove damaging to the embryos. If it was a matter of freezing week-old pig embryos, then clearly it would be essential to have a transfer technique capable of dealing with them.

In Korea, Kim and Oguri (1990) examined various factors affecting pregnancy rates and litter size after surgical transfers. They recorded pregnancy rates of 75% and 90% and litter sizes of 5.3 and 6.1 after transfers to recipients at 3 and 4 days after oestrus, respectively. For transfer of embryos of 5 and 6 days of age, they recorded pregnancy rates of 66.7% and 100% and litter sizes of 6.5 and 5.6, respectively. In the former East Germany, Brussow (1990) reported that embryo survival rate was significantly influenced by the number of embryos transferred; it was 64% after the transfer of 12–15 embryos and 45% after transfer of 16–20 embryos. In the former Yugoslavia, Kosarcic *et al.* (1992) recorded a pregnancy rate of 58% and a litter size of 8.9, after transferring 682 embryos to 52 recipients.

Minimum number of embryos required

One fact that should be borne in mind when talking about pig embryo transfer is that pregnancy may not be maintained unless the recipient pig receives at least four embryos (Polge *et al.*, 1966). Presumably, this is a matter of inadequate luteotrophic effect unless a certain minimum number of embryos is present in the uterus.

8.5.3. Non-surgical embryo transfer

As with sheep and goats, one factor limiting progress in the commercial application of pig embryo transfer (ET) on farms has been the necessity to employ surgical procedures for the transfer of embryos. Polge and Day (1968) showed that it was possible to achieve a pregnancy after a non-surgical approach, but the success rate was very low (1/32 attempts). This was not unexpected in view of the anatomy of the sow's cervix in which the small openings of the cervical rings, which are located eccentrically in the caudal half of the cervix, make it very difficult to introduce a catheter into the uterus. More recently, Sims and First (1987) transferred 7-day embryos on days 6 and 7 of the oestrous cycle, using a sterilized pig insemination spirette. In both of these reports, embryos were recovered during pregnancy and no piglets were born. Success was reported by Reichenbach *et al.* (1993) in Germany, in the form of live piglets, but the pregnancy rate was low (10%); the authors concluded that the method described was suitable for further evaluation.

In Missouri, Galvin *et al.* (1994) reported the birth of live piglets from gilts following non-surgical embryo transfer with a pregnancy rate and litter size of 63% and 7.1 with surgical and 21.7% and 4.3 with non-surgical transfers, respectively. Non-surgical recipients were sedated 15 minutes before transfer and a disposable insemination spirette manipulated into the cervix; embryos were expelled from a tomcat catheter into the reproductive tract. The Missouri workers speculated that the lower litter size after non-surgical transfers may have been due to the placement of embryos in the uterus; this was at the uterine tip with surgical and the cervical end with non-surgical transfers.

In Germany, Modl (1994), reporting on various experiments relating to non-surgical embryo transfer in pigs, recorded a pregnancy rate of 9.7% in recipients that received embryos in the uterine body compared with 43% for embryos placed in the caudal part of the uterine horn. A subsequent study by Wallenhorst and Holtz (1995) in the same country also examined the effect of uterine site on embryo survival rate (Table 8.3); it was evident that the uterine body was an unsuitable site and that the caudal quarter of the uterine horn was less suitable than the middle of the uterine horn.

In an evaluation of a transcervical embryo transfer procedure in trials reported from the Netherlands (Hazeleger and Kemp, 1994), for 21 sows in one trial, the farrowing rate was 33%. The authors recorded that pregnancy rates depended on the developmental stages of the transferred embryos; pregnancy rate was 55% with blastocysts and 10% with morulae. Hazeleger *et*

Table 8.3. Embryo survival rate in relation to different uterine sites after surgical transfer of pig embryos (from Wallenhorst and Holtz, 1995).

	Uterine site		
	Middle of horn	Caudal quarter of horn	Body
No. of recipients	16	16	17
Recipients pregnant (%)	88[a]	81[a]	12[b]
Fetuses per pregnancy ($x \pm$ SD)	8.2 ± 2.4[d]	5.6 ± 2.7[e]	4 and 5
Range	7 to 14	1 to 9	4 and 5
Embryo survival (%)	41[a]	28[b]	3[c]

[a,b,c]Means differ ($P < 0.05$, Chi^2-test).
[d,e]Means differ ($P < 0.05$, Scheffé-test).

al. (1995a) described work in which 13–20 embryos were transcervically transferred to recipient gilts without sedation; three of four synchronous and none of seven asynchronous transfers resulted in pregnancies. The authors concluded that exact synchrony may be of greater importance with non-surgical than with surgical transfers. A further report (Hazeleger *et al.*, 1995b) recorded that results of non-surgical transfers were best when recipients ovulated less than 12 h after the donor pigs.

Type of equipment for non-surgical transfers

A set of embryo transfer instruments has been described which enabled pig embryos (four-cell to blastocyst stages) to be delivered into the lumen of one uterine horn of recipient gilts (Li *et al.*, 1995). In small-scale experiments, a pregnancy rate of 31.3% and a litter size of 6.8 were recorded. The transfer instruments involve four parts: (i) a modified commercial AI spirette which is used to produce a cervical lock; (ii) stainless steel tubing with a curved tip that can be manipulated over the folds of the cervix; (iii) a testing bar to ensure that the tip of part (ii) reaches the bifurcation of the uterus; and (iv) a disposable tubing complex for embryo delivery. According to the report, a technician can easily learn to operate the instruments.

8.6. *In Vitro* Production of Pig Embryos

The development of *in vitro* maturation and *in vitro* fertilization techniques in pigs is of considerable importance because of the possible application of this technology in different areas of research. In contrast to results in the cow and the human, however, experiences to date with *in vitro* fertilization (IVF) in pigs have been much less successful. Limited evidence of fertilization by IVF in pigs was first reported by Japanese workers (Nagai *et al.*, 1984). Subsequently, live births were reported from Cambridge after the *in vitro* fertilization of ovulated pig oocytes (Cheng *et al.*, 1986). It was the work of Mattioli *et al.* (1989) in Italy, however, that first convincingly demonstrated that pig oocytes, matured and fertilized *in vitro* (IVM)/IVF), could undergo normal embryonic development. They reported blastocyst formation, establishment of pregnancies and the birth of live piglets.

8.6.1. Recovery and maturation of oocytes

Studies with the ovaries of slaughtered gilts, by Stancic *et al.* (1992) in the former Yugoslavia, showed the recovery of 7.9 and 45.3 oocytes per ovary for aspiration and dissection procedures, respectively. Although the collection of oocytes by dissection of follicles is very labour intensive in comparison with aspiration, it should ensure that oocytes are only recovered from non-atretic follicles. After collection, the selection of good quality cumulus–oocyte complexes (C-O-Cs) is necessary; it is important that the C-O-Cs selected should have compact and multiple cell layers.

In the maturation of the pig oocyte, the crucial importance of the interaction between somatic cells and the oocyte has been emphasized by many workers. In early studies, only maturation culture systems that involved the presence of abundant follicle wall components were able to create the appropriate conditions for competent oocyte maturation (Mattioli *et al.*, 1988). The presence of intact follicle shells in a non-static culture system appears to have a beneficial effect on the maturation of pig oocytes; the value of the flux system is believed to be in preventing premature differentiation of follicular cells (Nagai *et al.*, 1993). Other factors influencing the resumption of meiotic maturation of porcine oocytes have been examined in many studies.

In terms of the need to add hormones to the maturation medium, Nagai *et al.* (1993) found evidence that gonadotrophins and/or oestradiol could be beneficial. A report by Funahashi and Day (1993a) produced evidence of a high rate of male pronucleus formation in pig oocytes cultured for 20 h in medium with gonadotrophins (PMSG + hCG) and oestradiol, followed by 20 h culture in hormone-free medium. In studies reported by Yoshida *et al.* (1992), it was found that the addition of cysteine to the maturation medium was an important factor in enhancing the ability of pig oocytes to form male pronuclei after sperm penetration. In Canada, Singh *et al.* (1993) found that the addition of epidermal growth factor (EGF) to the maturation medium significantly increased the incidence of germinal vesicle breakdown.

In Australia, Grupen *et al.* (1996) examined the polyspermy problem in pig IVF in relation to the progression of meiosis in porcine oocytes matured *in vitro*. They noted that pig IVM systems have usually been based on a 40 h period and considered the possibility that this might be too long. They showed that reducing the maturation interval from 44 to 36 h significantly reduced the incidence of polyspermy from 34% to 5%. It was concluded that the problem of polyspermy may be due to the ageing of meiotically mature oocytes prior to fertilization.

In a review of progress, Nagai (1994) noted that IVM of pig oocytes was possible in a simple medium (mTALP) supplemented with pig follicular fluid and cysteine. The same author concluded that the success rate with IVM-IVF in the pig is low and there are obviously many steps in the procedures used which do not yet match up with what occurs *in vivo*.

8.6.2. Oocyte recovery from live pigs

In cattle, it is now possible to produce embryos from oocytes recovered by follicle aspiration, using either endoscopy or the ultrasound-guided transvaginal approach. Oocyte aspiration was investigated in gilts by follicular puncture via laparoscopy by Brussow and Ratky (1994a,b) in Germany, who reported that follicular and oocyte development was not influenced by repeated follicular puncture. Brussow *et al.* (1995) recovered porcine oocytes by endoscopy after oestrous synchronization with altrenogest, followed by

Table 8.4. Morphology of cumulus–oocyte complexes at different times before and after hCG administration (from Brussow *et al.*, 1995).

Time after hCG (h)	Number of C-O-C (Σ *n*)	Compact (%)	Slight compact (%)	Corona radiata (%)	Expanded (%)	Denuded (%)
−2	85	42.3	30.6[a]	18.8	0	8.2
10	47	25.5	31.9[a]	6.4	25.5[c]	10.6
22	60	0	5.0[b]	3.3	86.7[d]	5.0
34	59	0	0	1.7	98.3[d,e]	0

[a:b, c:d] $P < 0.01$; [d:e] $P < 0.5$.

PMSG and with hCG given after a further 80 h; follicles were aspirated at different intervals after hCG (Table 8.4). The results showed a high proportion of immature oocytes (germinal vesicle stage) 2 h before and 10 h after hCG; the number of mature oocytes tended to increase thereafter according to the time of recovery.

8.6.3. Capacitation of boar sperm

In pigs, sperm preparation and capacitation as employed by Mattioli *et al.* (1989) involved prolonged incubation (24–48 h) in a commercial semen diluent, with subsequent washing and use of Percoll layers to provide a highly motile sperm fraction. This was a modification of procedure previously reported by Cheng *et al.* (1986) at Cambridge. In Korea, Park *et al.* (1990) suggested that storage of boar semen at 20°C for 10–12 h and preincubation at a high sperm concentration was a suitable method of inducing capacitation of pig sperm. Other workers examined combination treatments with ionophore A23187, heparin and caffeine (Wang *et al.*, 1991; Lei *et al.*, 1992). Yoshida *et al.* (1993) reported the birth of piglets that had been produced by the IVF of *in vitro* matured oocytes using freshly ejaculated sperm in modified Tyrode's medium without preincubation. The capacitation procedures employed with boar sperm have generally been more complex and tedious than those used with cattle. This probably reflects the lack of knowledge about the capacitation requirements of the species.

8.6.4. *In vitro fertilization*

In a review of some problems facing the IVF of pig oocytes, Hunter (1990) suggested concentrating attention on the factors contributing to the oviductal microenvironment during *in vivo* fertilization in pigs. It is well recognized that one of the most acute problems in pig IVF is that of polyspermy. Cheng *et al.*

(1986) found that oocytes exposed to pig sperm for more than 8 h showed a significantly increased incidence of polyspermy. Mattioli *et al.* (1989) found that removal of adherent sperm from the zonae pellucidae 6–8 h after the start of gamete co-incubation reduced the incidence of multiple sperm penetration. Although other workers have followed this practice, polyspermy has remained a serious problem.

It would appear that there is a marked species difference between pigs and ruminants in the incidence of polyspermy after *in vitro* fertilization. According to Coy *et al.* (1993), this may be due to the blockage to polyspermy being physiologically slower in pigs than in cows and because exocytosis and dispersal of the cortical granule contents during IVF is more normal than in pigs; these Spanish workers used a gamete co-incubation period of 4 h and suggested that this may be optimal for the pig.

Funahashi and Day (1993b) showed that prefertilization incubation of boar sperm in suitable concentrations of pig follicular fluid could reduce the incidence of polyspermy in a pig IVF system. They suggested that follicular fluid, released by the sow ovary at ovulation, may assist in preventing polyspermy *in vivo*.

A further problem in pig IVF relates to the quality of available sperm. The difficulty of freezing boar sperm effectively has meant that fresh ejaculated semen has usually been employed in pig IVF. It is known that large variations exist among boars in the ability of their sperm to penetrate oocytes; the polyspermy problem is also greater with some boars than with others (Xu *et al.*, 1995). In Canada, Foxcroft *et al.* (1995) showed that sperm penetration, monospermy and male pronuclear formation were all significantly affected by boars.

The storage of boar semen in the fresh state, when this is over a period of several days, may also be a consideration. The fertilizing ability of boar semen stored in Kiev diluent (Hamano *et al.*, 1990) decreased from 98–100% at day 0/1 to 0% at day 5. Large variations in fertilization rate have been evident with frozen–thawed sperm (Nagai *et al.*, 1988). Using the zona-free hamster oocyte test, Clarke and Johnson (1987) showed that penetration by frozen–thawed boar sperm, separated by a swim-up procedure, was significantly decreased in comparison with penetration by sperm from fresh or liquid-stored ejaculates. To examine the effect of different sperm preparation methods Zheng *et al.* (1992) froze boar semen and subsequently used fractionation techniques to isolate highly motile sperm.

8.6.5. *In vitro* culture of early embryos

The methods employed for the *in vitro* culture of pig embryos produced by *in vitro* maturation and fertilization of follicular oocytes are essentially the same as those previously dealt with previously (see section 8.4). It is thus possible to produce pig embryos in the laboratory which are suitable for transfer. Yoshida *et al.* (1993), for example, collected oocytes from prepubertal gilts slaughtered at a local abattoir, matured and fertilized them and allowed the embryos to develop to the two- to four-cell stage in modified Krebs Ringer medium. The embryos were transferred to a recipient gilt which duly farrowed three healthy piglets after a normal gestation period. The effectiveness of their procedures, however, left much room for improvement, and the same can be said of pig IVF efforts in general.

A report by Laurincik *et al.* (1994) concluded that under their IVF conditions, oocytes matured *in vitro* displayed lower penetration and cleavage rates and asynchronous pronucleus development, as well as delayed cleavage, in comparison with oocytes matured *in vivo*. According to Rath *et al.* (1995), it is still preferable to transfer porcine embryos derived from *in vitro* fertilization at an early stage of development and not to culture them to the blastocyst stage before transfer.

8.6.6. Gamete intrafallopian transfer in pigs

In human-assisted reproduction, gamete intrafallopian transfer (GIFT) has been developed as an alternative to *in vitro* fertilization and embryo transfer. Chi *et al.* (1989), in Japan, examined the possibility of GIFT as an alternative to conventional embryo transfer in pigs. Oocytes were transferred with ejaculated sperm into three recipient gilts, two of which became pregnant and deliverd three piglets. An account of a GIFT procedure that resulted in the production of normal young in pigs was also given by Rath *et al.* (1994a,b) in Germany. These workers suggest that the establishment of a reliable GIFT technique might be useful when using gender preselected semen. They point to the evidence that separation of X- and Y-chromosome-bearing boar sperm is possible by use of flow cytometry (Johnson, 1991a,b).

8.7. Gene Manipulation, Sexing and Cloning in Pigs

The ability to mature, fertilize and develop pig oocytes *in vitro* is likely to be of great value in providing a supply of porcine oocytes and embryos for use in many areas of basic research as well as in work aimed at commercial application. Kubisch *et al.* (1995), for example, have demonstrated that such *in vitro* developed embryos could be used effectively in gene transfer research.

8.7.1. Production of transgenic pigs

In the past, genetic improvement in pigs has come about largely by way of testing, selecting and crossbreeding programmes. The future, however, holds the promise of exciting new methods whereby gene manipulation may be used to greatly accelerate breeding progress. Methods of gene manipulation in pigs involve the isolation, transfer and control of genes. The method first used for gene transfer involved the injection of the DNA into pronuclear oocytes. Although the first transgenic modifications of pigs was reported more than a decade ago (Hammer *et al.*, 1985; Miller *et al.*, 1986), research in this area would obviously be facilitated by an abundant supply of zygotes produced by *in vitro* techniques. The collection of ovulated one-cell porcine embryos from gilts has been detailed by Williams *et al.* (1990). However, microinjection of DNA is known to be detrimental to early zygote development and may contribute to the problem of low pregnancy rates experienced with transgenic embryos (Williams *et al.*, 1992); a ready supply of zygotes produced *in vitro* might assist in resolving such problems. In the production of transgenics, there may well be areas relevant to human health and welfare, such as the germ-line modification of pig organs for human transplantation, which may be facilitated by a ready supply of laboratory produced embryos.

It should be noted that the success rate has been very low (1%) in gene transfer attempts by way of the microinjection of DNA into the pronuclear oocyte. An alternative method is to use retroviruses containing the desired genes, which 'infect' the recipient pigs. A third approach, and one which is likely to be the most promising, is the use of embryonic stem cells, which may be genetically modified and then fused with a pig oocyte, using cloning technology. This could provide embryos that are then transferred to recipients to produce transgenic piglets. The greatest challenge in gene manipulation research lies in determining the control mechanisms for gene expression; only by this means will it be possible to introduce genes so that they give rise to specific and predicted responses.

8.7.2. Sexing in pigs

Some authors have reported significant differences in the sex ratio in pigs arising from the nature of their diet. In Japan, for example, Toriumi *et al.* (1993) reported on the effect of a Ca/P-enriched diet (2.15% Ca and 1.28% P) on the secondary sex ratio (that shown in piglets born); the male:female ratio for sows receiving a standard and enriched diet was 0.89 and 1.32, respectively.

Kato *et al.* (1987) described a method of sexing the pig embryo without adversely affecting its subsequent development which involved chromosomal analysis of trophectoderm cells taken from expanding blastocysts. The possibility of using sexed semen in pig IVF has received attention in some studies from Germany (Rath *et al.*, 1993, 1995b). Flow cytometry has been shown to be a reliable method of separating X- from Y-bearing sperm in cattle, humans and several other species, including the pig. The transfer of embryos fertilized by such sperm resulted in two recipient pigs becoming pregnant and delivering ten healthy piglets, all of which were of the predicted (female) sex (Rath *et al.*, 1995b).

8.7.3. Embryo splitting and cloning in pigs

The interest in embryo splitting in pigs stems from the fact that monozygotic twin piglets can serve as excellent models in various areas of medical and biological research. Embryo splitting technology has been the subject of many reports in cattle and with a smaller number of papers dealing with sheep and goats; few studies have been reported on embryo bisection in pigs. Two sets of identical twin pigs were described by Cambridge workers in the early 1980s (Willadsen, 1982; Polge, 1985); other reports dealt with live piglets resulting from split embryos but with no twin sets (Rorie *et al.*, 1985; Nagashima *et al.*, 1988).

Results of investigations reported by Reichelt and Niemann (1994) were more promising, showing that pig morulae and blastocysts could be bisected relatively easily and yield large numbers of viable demi-embryos. This was in contrast to findings by Robl and First (1985). The German workers showed that after transferring a sufficient number of demi-embryos per recipient, high pregnancy rates could be achieved, although with a reduced litter size. Identical twin piglets from the study underwent reciprocal skin and pancreas transplantations without showing evidence of any type of graft rejection. In other work in Germany, both Reichel *et al.* (1995) and Tao *et al.* (1995) concluded that pig embryos at the blastocyst stage give the maximum yield of viable demi-embryos after splitting. These findings were in agreement with Ash *et al.* (1989), who also concluded that pig embryos split at the blastocyst stage were more capable of surviving *in utero* than those split at earlier stages.

Cloning studies

A number of papers have reported studies relevant to cloning in pigs (Onishi and Youngs, 1993; Saito *et al.*, 1993; Stumpf *et al.*, 1993; Terlouw *et al.*, 1993; Parry and Prather, 1994; Gerfen and Wheeler, 1995). Much of the interest in this area lies in the possibility of developing embryonic stem cell-lines which may be used in the production of transgenic pigs, via cloning by nuclear transfer. The plan here is to transfer the genetic material into the genome of an embryonic stem cell and then use the cell in nuclear transfer.

8.8. References

Allen, R.L. and Wright, R.W.Jr (1984) In vitro development of porcine embryosin coculture with endometrial cell monolayers or culture supernatants. *Journal of Animal Science* 59, 1657–1660.

Arav, A. (1992) Vitrification of oocytes and embryos. In: Lauria, A. and Gandolfi, F. (eds) *Embryonic Development and Manipulation in Animal Production*. Portland Press, London, pp. 255–264.

Arav, A., Rubinsky, B., Fletcher, G. and Seren, E. (1993) Cryogenic protection of oocytes with antifreeze proteins. *Molecular Reproduction and Development* 36, 488–493.

Archibong, A.E., Petters, R.M. and Johnson, B.H. (1989) Development of porcine embryos from one- and two-cell stages to blastocysts in culture medium supplemented with porcine oviductal fluid. *Biology of Reproduction* 41, 1076–1083.

Ash, K., Anderson, G.B., BonDurant, R.H., Pashen, R.L., Parker, K.M. and Berger, T. (1989) Competition between split and nonmanipulated embryos in the production of identical piglets. *Theriogenology* 31, 903–910.

Averill, R.L.W., Adams, C.E. and Rowson, L.E.A. (1955) Transfer of mammalian ova between species. *Nature* 176, 167–170.

Baker, R.D. (1979) Embryo recovery from prepubertal gilts. *Theriogenology* 11, 91.

Baker, R.D. and Coggins, E.G. (1968) Control of ovulation rate and fertilization in prepubertal gilts. *Journal of Animal Science* 27, 1607–1610.

Bazer, F.W., Thatcher, W.W., Martinat-Botte, F. and Terqui, M. (1988) Conceptus development in Large White and prolific Chinese Meishan pigs. *Journal of Reproduction and Fertility* 84, 37–42.

Beckmann, L.S. and Day, B.N. (1993) Effects of media NaCl concentration and osmolarity on the culture of early-stage porcine embryos and the viability of embryos cultured in a selected superior medium. *Theriogenology* 39, 611–622.

Biggers, J.D., Gwatkin, R.B.L. and Brinster, R.L. (1962) Development of mouse embryos in organ cultures of fallopian tubes on a chemically defined medium. *Nature* 194, 747–749.

Bolamba, D. and Sirard, M-A. (1995) *In vitro* development of *in vivo* fertilized zygotes from superovulated gilts treated with a progesterone agonist (Regumate) or a prostaglandin agonist (Planate). *Biology of Reproduction* 52 (Suppl. 1), 176.

Brussow, K.P. (1990) Results of embryo transfer into the oviduct or uterus in pigs. *Monatshefte fur Veterinarmedizin* 45, 562–565.

Brussow, K.P. and Ratky, J. (1994a) Laparoscopical ovum pick-up in swine. *Proceedings of the 10th Meeting European Embryo Transfer Association* (Lyon), 160.

Brussow, K.P. and Ratky, J. (1994b) Repeated laparoscopical follicular puncture and oocyte aspiration in swine. *Reproduction in Domestic Animals* 29, 494–502.

Brussow, K.P., Torner, H., Ratky, J. and Alm, H. (1995) Endoscopical ovum pick up in gilts depending on follicular and oocyte maturation. *Proceedings of the 11th Meeting European Embryo Transfer Association* (Hanover), p. 140.

Cameron, R.D.A., Durack, M., Fogarty, R., Putra, D.K.H. and McVeigh, J. (1989) Practical experience with commercial embryo transfer in pigs. *Australian Veterinary Journal* 66, 314–318.

Chen, L.R. and Wu, M.C. (1991) Procedures for surgical collection of porcine embryos and uniformity of embryos. *Journal of Chinese Society of Animal Science* 20, 175–187.

Cheng, W.T.K., Moor, R.M. and Polge, C. (1986) In vitro fertilization of pig and sheep oocytes matured in vivo and in vitro. *Theriogenology* 25, 146.

Chi, C-H., Nagashima, H., Kato, Y., Mori, Y. and Hoshino, K. (1989) Normal development of follicular oocytes collected from prepubertal gilts after gamete intrafallopian transfer. *Japanese Journal of Animal Reproduction* 35, 217–221.

Clarke, R.N. and Johnson, L.A. (1987) Effect of liquid storage and cryopreservation of boar spermatozoa on acrosomal integrity and penetration of zona-free hamster ova *in vitro. Gamete Research* 16, 193–204.

Concar, D. (1994) The organ factory of the future? *New Scientist* 142, 24–29.

Coy, P., Martinez, E., Ruiz, S., Vazquez, J.M., Roca, J., Matas, C. and Pellicer, M.T. (1993) In vitro fertilization of pig oocytes after different coincubation intervals. *Theriogenology* 39, 1201–1208.

Curnock, R.M., Day, B.N. and Dziuk, P.J. (1976) Embryo transfer in pigs: a method for introducing genetic material into primary specific-pathogen-free herds. *American Journal of Veterinary Research* 37, 97–98.

Davis, D.L. (1985) Culture and storage of pig embryos. *Journal of Reproduction and Fertility* Suppl. 33, 115–124.

Davis, D.L. and Day, B.N. (1978) Cleavage and blastocyst formation by pig eggs *in vitro*. *Journal of Animal Science* 46, 1043–1053.

Day, B.N. (1979) Embryo transfer in swine. *Theriogenology* 11, 27–31.

Didion, B.A., Pomp, D., Martin, M.J. and Markert, C.L. (1990) Observations on the cooling and cryopreservation of pig oocytes at the germinal vesicle stage. *Journal of Animal Science* 68, 2803–2810.

Dobrinsky, J.R. (1993) Cryopreservation of porcine embryos: current status and perspectives. *Embryo Transfer Newsletter* 11, 13–18.

Dobrinsky, J.R. and Johnson, L.A. (1993) Effect of vitrification media on the *in vitro* development of porcine embryos. *Theriogenology* 39, 209.

Dobrinsky, J.R., Overstrom, E.W. and Johnson, L.A. (1994) Cytoskeletal analysis of vitrified porcine embryos. *Journal of Animal Science* 72 (Suppl. 1)/*Journal of Dairy Science* 77 (Suppl. 1), 82.

Dziuk, P.J. (1960) Frequency of spontaneous fragmentation of the ova of unbred gilts. *Proceedings Society of Experimental Biology* (New York) 103, 91–92.

Dziuk, P.J.and Day, B.N. (1977) Research applications of embryo transfer in pigs. In: Betteridge, K. (ed.) *Embryo Transfer in Farm Animals*, Monograph 16, Canadian Department of Agriculture, Ottawa, pp. 10–13.

Dziuk, P.J., Polge, G. and Rowson, L.E.A. (1964) Intrauterine migration and mixing of embryos in swine following egg transfer. *Journal of Animal Science* 23, 37–42.

Eroglu, A. (1993) Experimental studies on in vitro maturation of porcine oocytes. II. Effects of oestradiol-17β and progesterone. *Berliner und Munchener Tierarztliche Wochenschrift* 106, 157–159.

Feng, S., Zhang, Y., Li, S., Ma, Z., Wang, R. and Lu, D. (1991) Piglets from frozen (-20°C) embryos were born in China. *Theriogenology* 35, 199.

Foxcroft, G.R., Xu, X., Seth, P.C., Harbison, D.S. and Cheung, A.P. (1995) Semen dilution for assessment of boar ejaculate quality in pig IVM and IVF systems. *Theriogenology* 43, 212.

Funahashi, H. and Day, B.N. (1993a) Effects of the duration of exposure to supplemental hormones on cytoplasmic maturation of pig oocytes in vitro. *Journal of Reproduction and Fertility* 98, 177–185.

Funahashi, H. and Day, B.N. (1993b) Effects of follicular fluid at fertilization *in vitro* on sperm penetration in pig oocytes. *Journal of Reproduction and Fertility* 99, 97–103.

Galvin, J.M., Stewart, A.N.V. and Meredith, S. (1993) Higher sodium chloride concentration can induce a four-cell block in porcine embryos. *Theriogenology* 39, 224.

Galvin, J.M., Killian, D.B. and Stewart, A.N.V. (1994) A procedure for successful nonsurgical embryo transfer in swine. *Theriogenology* 41, 1279–1289.

Gerfen, R.W. and Wheeler, M.B. (1995) Isolation of embryonic cell-lines from porcine blastocysts. *Animal Biotechnology* 6, 1–14.

Grupen, C.G., Nagashima, H. and Nottle, M.B. (1996) Progression of meiosis in porcine oocytes matured in vitro and the polyspermy problem. *Theriogenology* 45, 190.

Guthrie, H.D., Henricks, D.M. and Handlin, D.L. (1974) Plasma hormone levels and fertility in pigs induced to superovulate with PMSG. *Journal of Reproduction and Fertility* 41, 361–370.

Gwatkin, R.B.L. and Biggers, J.D. (1963) Histology of mouse fallopian tubes maintained as organ cultures on a chemically defined medium. *International Journal of Fertility* 8, 453–457.

Hagen, D.R., Prather, R.S., Sims, M.M. and First, N.L. (1991) Development of one-cell porcine embryos to the blastocyst stage in simple media. *Journal of Animal Science* 69, 1147–1150.

Hajdu, M.A., Knight, J.W., Canseco, R.S., Krisher, R.L., Velander, W.H., Peason, R.E. and Gwazdauskas, F.C. (1994) Effect of culture conditions, donor age, and injection site on in vitro development of DNA microinjected porcine zygotes. *Journal of Animal Science* 72, 1299–1305.

Hamano, S., Toyoda, Y., Naito, K. and Fukuda, Y. (1990) In vitro fertilization of porcine oocytes with spermatozoa preserved in Kiev extender at 18°C. *Japanese Journal of Animal Reproduction* 36, 231–234.

Hammer, R.E., Pursel, V.G., Rexroad, C.E., Wall, R.J., Bolt, D.J., Ebert, K.M., Palmiter, R.D. and Brinster, R.L. (1985) Production of transgenic rabbits, sheep and pigs by microinjection. *Nature* 315, 680–685.

Hancock, J.L. (1961) Fertilization in the pig. *Journal of Reproduction and Fertility* 2, 307–331.

Hancock, J.L. and Hovell, G.J.R. (1962) Egg transfer in the sow. *Journal of Reproduction and Fertility* 4, 195–201.

Hayashi, S., Kobayashi, K., Mizuno, J., Saitoh, K. and Hirano, S. (1989) Development of pig fetuses from embryos frozen at −35°C. *Japanese Journal of Animal Reproduction* 35, 135–139.

Hayashi, S., Kobayashi, K., Mizuno, J. and Ohishi, R. (1994) Studies on embryo transfer in pigs. 1. The effects of repeated surgical embryo recovery from the same donors on ovarian function and embryo recovery. *Japanese Journal of Swine Science* 31, 8–14.

Hazeleger, W. and Kemp, B. (1994) Farrowing rate and litter size after transcervical embryo transfer in sows. *Reproduction in Domestic Animals* 29, 481–487.

Hazeleger, W., Mellen, J.Van Der and Lende, T.Van Der (1989) A method for transcervical embryo collection in the pig. *Theriogenology* 32, 727–734.

Hazeleger, W., Jenneskens, P. and Kemp, B. (1995a) Non-surgical transfer of porcine blastocysts. *Theriogenology* 43, 232.

Hazeleger, W., Peeters, I. and Kemp, B. (1995b) Asynchronous non-surgical transfer of porcine blastocysts. *Proceeding of the 11th Meeting European Embryo Transfer Association* (Hanover), 188.

Hunter, R.H.F. (1964) Superovulation and fertility in the pig. *Animal Production* 6, 189–194.

Hunter, R.H.F. (1974) Chronological and cytological details of fertilization and early embryonic development in the domestic pig, *Sus scrofa*. *Anatomical Record* 178, 169–186.

Hunter, R.H.F. (1988) *The Fallopian Tubes*. Springer-Verlag, London, 191 pp.

Hunter, R.H.F. (1990) Fertilization of pig eggs in vivo and in vitro. *Journal of Reproduction and Fertility* Suppl. 40, 211–226.

Hunter, R.H.F., Polge, C. and Rowson, L.E.A. (1966) The recovery, transfer and survival of blastocysts in pigs. *Journal of Reproduction and Fertility* 14, 501–502.

Illera, M.J., Lorenzo, P., Illera, J.C., Portela, A. and Petters, R.M. (1992) A simple medium supports in vitro development of one-cell embryos from miniature pigs to the blastocyst stage. *Theriogenology* 37, 225.

Iwasaki, S., Mizuno, J., Kobayashi, K., Yoshikane, Y. and Hayashi, T. (1994) Changes in morphology and cell number of inner cell mass of porcine blastocysts during freezing. *Theriogenology* 42, 841–848.

James, J.E. and Reeser, P.D. (1979) Embryo recovery in swine. *Theriogenology* 11, 47–50.

James, J.E., Reeser, P.D., Davis, D.L., Straiton, E.C., Talbot, A.C. and Polge, C. (1980) Culture and long-distance shipment of swine embryos. *Theriogenology* 14, 463–469.

Jarrell, V.L., Beckmann, L.S., Cantley, T.C., Rieke, A.R. and Day, B.N. (1990) The effect of exposure to an asynchronous uterus on development of day 4 swine embryos. *Journal of Animal Science* 68 (Suppl. 1), 429.

Jarrell, V.L., Day, B.N. and Prather, R.S. (1991) The transition from maternal to zygotic control of development occurs during the 4-cell stage in the domestic pig, *Sus scrofa*: quantitative and qualitative aspects of protein synthesis. *Biology of Reproduction* 44, 62–68.

Johnson, L.A. (1991a) Verified sex pre-selection in farm animals. *Proceedings of the 2nd International Conference on Boar Semen Preservation* (Beltsville), pp. 213–219.

Johnson, L.A. (1991b) Sex preselection in swine: altered sex ratios in offspring following surgical insemination of flow-sorted X- and Y-bearing sperm. *Reproduction in Domestic Animals* 26, 309–314.

Kameyama, K., Takedomi, T., Itakura, H. and Onihara, T. (1990) Effect of lecithin on cryopreservation of porcine embryos. *Proceedings of the 78th Annual Conference of Japanese Society of Animal Reproduction* (Niigata), 22.

Kang, S.R. and Wu, M.C. (1991) Induction of superovulation in gilts with gonadotropins. *Journal of Taiwan Livestock Research* 24, 33–40.

Kashiwazaki, N., Ohtani, S., Miyamoto, K. and Ogawa, S. (1991) Production of normal piglets from hatched blastocysts frozen at −196°C. *Veterinary Record* 16, 256–257.

Kato, Y., Nagashima, H., Shibata, K., Ishiwata, M. and Ogawa, S. (1987) Microsurgical bisection of expanding blastocysts for the sexing of pig embryos. *Journal of Animal Reproduction* 33, 151–154.

Kim, H.S. and Oguri, N. (1990) Studies on embryo transfer in pigs. II. Factors affecting pregnancy rate and litter size. *Korean Journal of Animal Sciences* 32, 125–130.

Kim, H.S., Kojima, T. and Soma, T. (1990) Studies on embryo transfer in the pig. I. Effect of days and repeat collection on embryo production. *Korean Journal of Animal Sciences* 32, 9–14.

Knox, R.V. and Zimmerman, D.R. (1995) Development and evaluation of a treatment regimen for superovulation and embryo survival in pigs. *Theriogenology* 43, 251.

Kosarcic, D., Veselinovic, S., Veselinovic, S., Miljkovic, V., Mrvos, G., Kuzmanov, D. and Kosarcic, S. (1992) Embryo transfer in pigs in Vojvodina. *Veterinarski Glasnik* 46, 409–412.

Krisher, R.L., Petters, R.M., Johnson, B.H., Bavister, B.D. and Archibong, A.E. (1989) Development of porcine embryos from the one-cell stage to blastocyst in mouse oviducts maintained in organ culture. *Journal of Experimental Zoology* 249, 235–239.

Krisher, R.L., Petters, R.M. and Johnson, B.H. (1989) Effect of oviductal condition on the development of one-cell porcine embryos in mouse or rat oviducts maintained in organ culture. *Theriogenology* 32, 885–892.

Kronnie, G. and Boer, P. (1993) Embryonic development in the pig up to the 64-cell stage, with reference to DNA replication and cell cycle times from the third cleavage division. *Theriogenology* 39, 919–928.

Kubicova, E., Oberfranc, M., Grafenau, P., Pivko, J. and Laurincil, J. (1993) Superovulation of gilts with Folligon. *Journal of Farm Animal Science* 26, 17–20.

Kubisch, H-M., Larson, M.A., Funahashi, H., Day, B.N. and Roberts, R.M. (1995) Pronuclear visibility, development and transgene expression in IVM/IVF-derived porcine embryos. *Theriogenology* 44, 391–401.

Kvasnicki, A.V. (1951) Interbreed ova transplantation. *Sovetsk.Zooteh.* pp. 36–42. *Animal Breeding Abstracts* 19, 224.

Laurincik, J., Rath, D. and Niemann, H. (1994) Differences in pronucleus formation and first cleavage following *in vitro* fertilization between pig oocytes matured *in vivo* and *in vitro*. *Journal of Reproduction and Fertility* 102, 277–284.

Lei, C., Wang, Z.K., Wei, P.H., Wang, J.Z., Kong, M.Q. and Wu, T.G. (1992) Comparison of different *in vitro* capacitation methods for pig sperm assessed by penetration of zona-free hamster eggs. *Theriogenology* 37, 243.

Li, J., Rieke, A. and Day, B.N. (1995) Technical breakthrough in non-surgical porcine embryo transfer. *Biology of Reproduction* 52 (Suppl. 1), 127.

Martin, P.A. (1986) Embryo transfer in swine. In: Morrow, D.A. (ed.) *Current Therapy in Theriogenology*. W.B. Saunders, Philadelphia, pp. 66–69.

Mattioli, M., Galeati, G., Bacci, M.L. and Seren, E. (1988) Follicular factors influence oocyte fertilizability by modulating the intercellular cooperation between cumulus cells and oocyte. *Gamete Research* 21, 223–232.

Mattioli, M., Bacci, M.L., Galeati, G. and Seren, E. (1989) Developmental competence of pig oocytes matured and fertilized *in vitro*. *Theriogenology* 31, 1201–1207.

Menino, A.R.Jr and Wright, R.W.Jr (1982) Development of one cell porcine embryos in two culture systems. *Journal of Animal Science* 54, 583–588.

Miller, K.F., Bolt, D.J., Pursel, V.G., Hammer, R.E., Pinkert, C.A., Palmiter, R.D. and Brinster, R.L. (1986) Expression of human or bovine growth hormone gene with a mouse metallothionein-1 promoter in transgenic swine alters the secretion of porcine growth hormone and insulin-like growth factor-I. *Journal of Endocrinology* 120, 481–488.

Modl, J. (1994) Investigations on oestrus synchronization, superovulation, embryo freezing and non-surgical embryo transfer in pigs. Thesis, Ludwig-Maximilians-Univerasitat Munchen, Germany, 128 pp.

Murphy, M. (1984) Surgical swap. *Hog Farm Management* (June issue), 7–12.

Nagai, T. (1994) Current status and perspectives in IVM-IVF of porcine oocytes. *Theriogenology* 41, 73–78.

Nagai, T., Niwa, K. and Iritani, A. (1984) Effect of sperm concentration during preincubation in a defined medium on fertilization *in vitro* of pig follicular oocytes. *Journal of Reproduction and Fertility* 70, 271–275.

Nagai, T., Ding, J.C. and Moor, R.M. (1988) Factors affecting sperm penetration and pronuclear formation in pig oocytes matured *in vitro*. *Journal of Reproduction and Fertility*, Abstract Series no. 2, Abs. 68.

Nagai, T., Ding, J. and Moor, R.M. (1993) Effect of steroidogenesis on maturation and fertilization *in vitro* of pig oocytes. *Journal of Experimental Zoology* 266, 146–151.

Nagashima, H., Katoh, Y., Shibata, K. and Ogawa, S. (1988) Production of normal piglets from microsurgically split morulae and blastocysts. *Theriogenology* 29, 485–495.

Nagashima, H., Kato, Y. and Ogawa, S. (1989a) Microsurgical bisection of porcine morulae and blastocysts to produce monozygotic twin pregnancy. *Gamete Research* 23, 1–9.

Nagashima, H., Kato, Y., Yamakawa, H., Matsumoto, T. and Ogawa, S. (1989b) Changes in freezing tolerance of pig blastocysts in pre-hatching stage. *Japanese Journal of Animal Reproduction* 35, 130–134.

Nagashima, H., Yamakawa, H. and Niemann, H. (1992) Freezability of porcine blastocysts at different peri-hatching stages. *Theriogenology* 37, 839–850.

Nagashima, H., Kashawaszaki, N., Ashman, R., Grupen, C., Seamark, R.F. and Nottle, M. (1994) Recent advances in cryopreservation of porcine embryos. *Theriogenology* 41, 113–118.

Nagashima, H., Kashiwazaki, N., Ashman, R.J., Grupen, C.G. and Nottle, M.B. (1995a) Cryopreservation of porcine embryos. *Nature* 374, 416.

Nagashima, H., Kashiwazaki, N., Ashman, R.J. and Nottle, M.B. (1995b) Improved survival of porcine hatched blastocysts cryopreserved with glycerol and sucrose. *Journal of Reproduction and Development* 41, 165–170.

Nagashima, H., Kuwayama, M., Grupen, C.G., Ashman, R.J. and Nottle, M.B. (1996) Vitrification of porcine early cleavage stage embryos and oocytes after removal of cytoplasmic lipid droplets. *Theriogenology* 45, 180.

Naito, K., Fukuda, Y. and Ishibashi, I. (1989) Developmental ability of porcine ova matured in porcine follicular fluid *in vitro* and fertilized *in vitro*. *Theriogenology* 31, 1049–1057.

Niemann, H. (1985) Sensitivity of pig morulae to DMSO/PVP or glycerol treatment and cooling to 10°C. *Theriogenology* 23, 213.

Niemann, H. (1989) To South America in a thermos flask: first experiences of international transportation of pig embryos. *Tierzuchter* 41, 336–337.

Niimura, S. and Ishida, K. (1980) Histochemical observation of lipid droplets in mammalian eggs during the early development. *Japanese Journal of Animal Reproduction* 26, 46–49.

Onishi, A. and Youngs, C.R. (1993) Comparison of immunosurgery with calcium ionophore treatment for isolation of inner cell masses of porcine blastocysts. *Theriogenology* 39, 274.

Oxenreider, S.L. and Day, B.N. (1965) Transport and cleavage of ova in swine. *Journal of Animal Science* 24, 413–417.

Park, S.B., Park, H.K. and Iritani, A. (1990) *In vitro* fertilization in the pig. *Korean Journal of Animal Sciences* 32, 20–26.

Parry, T.W. and Prather, R.S. (1994) Protein reprogramming after nuclear transfer. *Journal of Animal Science* 72 (Suppl. 1)/*Journal of Dairy Science* 77 (Suppl. 1), 175.

Petters, R.M. and Reed, M.L. (1991) Addition of taurine or hypotaurine to culture medium improves development of one and two-cell pig embryos in vitro. *Theriogenology* 35, 253.

Phillipo, M. (1968) Superovulation in the pig. In: McLaren, A. (ed.) *Advances in Reproductive Physiology*, Vol. 3. Logos Press, London, pp. 147–166.

Pipkin, J.L., Prien, S.D., Pipkin, K.E. and Clark, J.R. (1991) Culture and co-culture of porcine embryos. *Journal of Animal Science* 69 (Suppl. 1), 417.

Plante, C., Pollard, J.W., Kobayashi, S. and Leibo, S.P. (1993) Chilling sensitivity of porcine morulae. *Theriogenology* 39, 285.

Polge, C. (1977a) *In vitro* fertilization and use of follicular oocytes. In: Betteridge, K.J. (ed.) *Embryo Transfer in Farm Animals*, Monograph No. 16, Canadian Department of Agriculture, Ottawa, p. 43.

Polge, C. (1977b) The freezing of mammalian embryos: perspectives and possibilities. In: Elliot, K. and Whelan, K. (eds) *The Freezing of Mammalian Embryos*. Elsevier, Amsterdam, pp. 3–18.

Polge, C. (1980) Embryo transplantation: a place in the future of pig production? *Pig Farming* 28, 75–79.

Polge, C. (1982) Embryo transplantation and preservation. In: Cole, D.J.A. and Foxcroft, G.R. (eds) *Control of Pig Production.* Butterworths, London, pp. 277–291.

Polge, C. (1985) How does embryo manipulation fit into present and future pig production? *Journal of Reproduction and Fertility* Suppl. 33, 93–100.

Polge, C. and Day, B.N. (1968) Pregnancy following non-surgical egg transfer in pigs. *Veterinary Record* 82, 712.

Polge, C. and Dziuk, P.J. (1970) Time of cessation of intrauterine migration of pig embryos. *Journal of Animal Science* 31, 565–567.

Polge, C., Rowson, L.E.A. and Chang, M.C. (1966) The effect of reducing the number of embryos during early stages of gestation on the maintenance of pregnancy in the pig. *Journal of Reproduction and Fertility* 12, 395–397.

Polge, C., Adams, C.E. and Baker, R.D. (1972) Development and survival of pig embryos in the rabbit oviduct. *Proceedings of the 7th International Congress on Animal Reproduction and AI* (Munich), pp. 513–517.

Polge, C., Wilmut, I. and Rowson, L.E.A. (1974) The low temperature preservation of cow, sheep and pig embryos. *Cryobiology* 11, 560.

Pomeroy, R.W. (1960) Infertility and neonatal mortality in the sow. Parts I to IV. *Journal of Agricultural Science, Cambridge* 54, 1–56.

Pope, G.E., Christenson, R.K., Zimmermann-Pope, V.A. and Day, B.N. (1972) Effect of number of embryos on embryonic survival in recipient gilts. *Journal of Animal Science* 35, 805–808.

Pope, G.E. and Day, B.N. (1977) Transfer of preimplantation pig embryos following in vitro culture for 24 to 48 hours. *Journal of Animal Science* 44, 1036–1040.

Pope, W.F., Lawyer, M.S., Nara, B.S. and First, N.L. (1986) Effect of asynchronous superinduction on embryo survival and range of blastocyst development in swine. *Biology of Reproduction* 35, 133–137.

Pope, W.F., Xie, S., Broermann, D.M. and Nephew, K.P. (1990) Causes and consequences of early embryonic diversity in pigs. *Journal of Reproduction and Fertility* Suppl. 40, 251–260.

Pursel, V.G. (1991) Production of transgenic swine. *Proceedings of the 2nd International Conference Boar Semen Preservation,* pp. 187–197.

Pursel, V.G. and Wall, R.J. (1995) Effects of transferred ova per recipient on production of transgenic swine. *Journal of Animal Science* 73 (Suppl. 1), 211.

Rath, D., Johnson, L.A. and Welch, G.R. (1993) In vitro culture of porcine embryos: development to blastocysts after in vitro fertilization (IVF) with flow cytometrically sorted and unsorted semen. *Theriogenology* 39, 293.

Rath, D., Johnson, L.A., Welch, G.R. and Niemann, H. (1994a) Successful gamete intrafallopian transfer (GIFT) in the porcine. *Theriogenology* 41, 1173–1179.

Rath, D., Niemann, H. and Johnson, L.A. (1994b) Gamete intrafallopian transfer (GIFT), an alternative to in vitro fertilization procedures for special applications. *Reproduction in Domestic Animals* 29, 349–351.

Rath, D., Niemann, H. and Torres, C.R.L. (1995) *In vitro* development to blastocysts or early porcine embryos produced *in vivo* or *in vitro. Theriogenology* 43, 913–926.

Rath, D., Johnson, L.A., Dobrinsky, J., Welch, G.R. and Niemann, H. (1995) Pregnancies following transfer of porcine oocytes fertilized in vitro with gender sorted semen. *Proceedings of the 11th Meeting of the European Embryo Transfer Association* (Hanover), p. 228.

Reed, M.L., Illera, M.J. and Petters, R.M. (1992) *In vitro* culture of pig embryos. *Theriogenology* 37, 95–109.
Reichelt, B. and Niemann, H. (1994) Generation of identical twin piglets following bisection of embryos at the morula and blastocyst stage. *Journal of Reproduction and Fertility* 100, 163–172.
Reichelt, B., Tao, T. and Niemann, H. (1995) Inner cell mass and trophoblast cells in demi- and intact porcine embryos. *Proceedings of the 11th Meeting of the European Embryo Transfer Association* (Hanover), p. 230.
Reichenbach, H.D., Modl, J. and Brem, G. (1993) Piglets born after transcervical transfer of embryos into recipient gilts. *Veterinary Record* 133, 36–39.
Robl, J.M. and First, N.L. (1985) Manipulation of gametes and embryos in the pig. *Journal of Reproduction and Fertility* Suppl. 33, 101–114.
Rorie, R.W., Voelkel, S.A., McFarland, C.W., Southern, L.L. and Godke, R.A. (1985) Micromanipulation of day-6 porcine embryos to produce split-embryo piglets. *Theriogenology* 23, 225.
Rubinsky, B., Arav, A. and Devries, A.L. (1991) Cryopreservation of oocytes using directional cooling and antifreeze glycoproteins. *Cryo-Letters* 12, 93–106.
Saito, S., Choi, Y.H. and Oguri, N. (1993) The parthenogenetic development of porcine oocytes matured in vitro after electrical ionophore A and ethanol activation. *Theriogenology* 39, 303.
Schlieper, B. and Holtz, W. (1986) Transfer of pig embryos collected by laparotomy or slaughter. *Animal Reproduction Science* 12, 109–114.
Schoenbeck, R.A., Peters, M.S., Rickords, L.F., Stumpf, T.T. and Prather, R.S. (1993) Timing of the transition from maternal to embryonic control in the 4-cell porcine embryo. *Theriogenology* 39, 305.
Sims, M.M. and First, N.L. (1987) Nonsurgical embryo transfer in swine. *Journal of Animal Science* 65 (Suppl. 1), 386.
Singh, B., Barbe, G.J. and Armstrong, D.T. (1993) Factors influencing resumption of meiotic maturation and cumulus expansion of porcine oocyte-cumulus cell complexes *in vitro*. *Molecular Reproduction and Development* 36, 113–119.
Singh, E.L. (1987) The disease control potential of embryos. *Theriogenology* 27, 9–20.
Smith, S., Schmidt, M., Purwantara, B. and Greve, T. (1993) Oviduct epithelial cell co-culture of early porcine embryos. *Acta Veterinaria Scandinavica* 33, 349–355.
Soede, N.M. and Kemp, B. (1993) The accessory sperm count is related to early embryonic diversity in pigs. *Theriogenology* 40, 1057–1064.
Springmann, K., Binder, S., Meier, E. and Brem, G. (1988) *Proceedings of the 11th International Congress on Animal Reproduction and AI* (Dublin), 2, Paper no. 196.
Sreenan, J.M. (1988) Embryo transfer: its uses and recent developments. *Veterinary Record* 122, 624–629.
Sreenan, J.M. and Scanlon, P.F. (1968) Continued cleavage of fertilized bovine ova in the rabbit. *Nature* 217, 867.
Stancic, B., Laurincik, J., Sahinovic, R., Kubavicova, E. and Oberfranc, M. (1992) The recovery, numbers and morphology of follicular oocytes in pigs. *Biotechnologija u Stocarstvu* 8(1–2), 23–31.
Stumpf, T.T., Terlouw, S.L., Funahashi, H., Prather, R.S. and Day, B.N. (1993) Processing of nuclei transplanted into *in vitro* matured porcine oocytes. *Theriogenology* 39, 322.
Takada, M., Yonezawa, N., Yoshizawa, M., Noguchi, S., Hatanaka, Y., Nagai, T., Kikuchi, K., Aoki, H. and Nakano, M. (1994) pH-sensitive dissociation and

association of β-N-acetylhexosaminidase from boar sperm acrosome. *Biology of Reproduction* 50, 860–868.

Tao, T., Reichelt, B. and Niemann, H. (1995) Ratio of inner cell mass and trophoblastic cells in demi- and intact pig embryos. *Journal of Reproduction and Fertility* 104, 251–258.

Terlouw, S.L., Stumpf, T.T., Funahashi, H., Prather, R.S. and Day, B.N. (1993) Pig oocyte activation and processing of transplanted nuclei. *Theriogenology* 39, 329.

Toriumi, H., Ohba, S., Kuwabara, Y., Takagi, K., Tsumagari, S. and Takeishi, M. (1993) Effects of high calcium and phosphorus administration on secondary sex ratio in pigs. *Journal of Reproduction and Development* 39, 73–77.

Veselinovic, S., Kosarcic, D., Veselinovic, S., Miljkovic, V., Mrvos, G., Kuzmanov, D., Ninkov, I., Murgaski, S., Stancic, B. and Pursel, V.G. (1991) Embryo transfer in swine. *World Review Animal Production* 26, 67–68.

Wallenhorst, S. and Holtz, W. (1995) Transfer of pig embryos to different uterine sites. *Proceedings of the 11th Meeting of the European Embryo Transfer Association* (Hanover), p. 256.

Wang, W.H., Niwa, K. and Okuda, K. (1991) *In vitro* penetration of pig oocytes matured in culture by frozen–thawed ejaculated spermatozoa. *Journal of Reproduction and Fertility* 93, 491–496.

Webel, S.K., Peters, J.B. and Anderson, L.L. (1970) Synchronous and asynchronous transfer of embryos in the pig. *Journal of Animal Science* 30, 565–568.

Weber, P.K. and Youngs, C.R. (1994) Investigation of cryoprotectant toxicity to porcine embryos. *Theriogenology* 41, 1291–1298.

White, K.L., Hehnke, K., Rickords, L.F., Southern, L.L. Thompson, D.L. and Wood, T.C. (1989) Early embryonic development in vitro by co-culture with oviductal epithelial cells in pigs. *Biology of Reproduction* 41, 425–430.

Whittingham, D.G. (1971) Survival of mouse embryos after freezing and thawing. *Nature* 233, 125–126.

Willadsen, S.M. (1982) Micromanipulation of embryos of the large domestic species. In: Adams, C.E. (ed.) *Mammalian Egg Transfer.* CRC Press, Boca Raton, pp. 185–210.

Williams, B.L., Knight, J.W., Canseco, R.S., Sparks, A.E.T., Gwazdauskas, F.C., Kornegay, E.T., Velander, W.H. and Wilkins, T.D. (1990) Comparison of synchronization procedures for the collection of one-cell porcine embryos. *Journal of Animal Science* 68 (Suppl. 1), 55.

Williams, B.L., Sparks, A.E.T., Canseco, R.S., Knight, J.W., Johnson, J.L., Velander, W.H., Page, R.L., Drohan, W.N., Young, J.M., Pearson, R.E., Wilkins, T.D. and Gwazdauskas, F.C. (1992) *In vitro* development of zygotes from prepubertal gilts after microinjection of DNA. *Journal of Animal Science* 70, 2207–2211.

Wilmut, I. (1972) The low temperature preservation of mammalian embryos. *Journal of Reproduction and Fertility* 31, 513–514.

Wollenberg, C., Wentz, I., Blum, B. and Holtz, W. (1990) Survival of pig embryos flushed from the reproductive tract immediately or two hours after slaughter of donors. *Journal of Animal Science* 68, 2023–2026.

Wrathall, A.E. (1984) Embryo transfer and disease transmission in swine. *Proceedings of the 10th International Congress on Animal Reproduction and AI* (Illinois), IV, IX 10–16.

Wrathall, A.E., Done, J.T., Stuart, P., Mitchell, D., Betteridge, K.J. and Randall, G.C.B. (1970) Successful intercontinental pig conceptus transfer. *Veterinary Record* 87, 226–268.

Wright, R.W.Jr (1977) Successful culture in vitro of swine embryos to the blastocyst stage. *Journal of Animal Science* 44, 854–858.

Xu, X., Kojima, T., Liu, T. and Ge, Y. (1994) The effect of heat shock on the tolerance to low temperatures in pig embryos. *Jiangsu Journal of Agricultural Sciences* 10, 42–45.

Xu, X., Ding, J., Seth, P.C., Harbison, D.S. and Foxcroft, G.R. (1995) *In vitro* fertilization of *in vitro* matured pig oocytes: effects of boar and ejaculate fraction. *Theriogenology* 43, 358.

Yamakawa, H., Nagasima, H., Kato, Y. and Tanioka, Y. (1990) Embryo collection from minature pigs and production of the piglets after transfer to common pig recipients. *Jikke Dobotsu, -Experimental Animals* 39, 413–415.

Yoshida, M., Ishigaki, K. and Pursel, V.G. (1992) Effect of maturation media on male pronucleus formation in pig oocytes matured *in vitro*. *Molecular Reproduction and Development* 31, 68–71.

Yoshida, M., Mizoguchi, Y., Ishigaki, K., Kojima, T. and Nagai, T. (1993) Birth of piglets derived from *in vitro* fertilization of pig oocytes matured *in vitro*. *Theriogenology* 39, 1303–1311.

Youngs, C.R., McGinnis, L.K. and Ford, S.P. (1991) *In vitro* development of Chinese Meishan pig embryos. *Serono Symposium on Preimplantation Embryo Development*, Abstract Volume, p. 174.

Youngs, C.R., Ford, S.P., McGinnis, L.K. and Anderson, L.H. (1993) Investigations into the control of litter size in swine. I. Comparative studies on in vitro development of Meishan and Yorkshire preimplantation embryos. *Journal of Animal Science* 71, 1561–1565.

Zheng, Y.S., Fiser, P. and Sirard, M.-A. (1992) The use of ejaculated boar semen after freezing in 2 or 6% glycerol for *in vitro* fertilization of porcine oocytes matured *in vitro*. *Theriogenology* 38, 1065–1075.

9 Breeding Pigs at Younger Ages

9.1. Introduction

Many of the piglets produced each year are farrowed by gilts and it follows from this that the age at which these females reach puberty is of considerable practical interest. In the USA, Tess *et al.* (1983) summarized the benefit of decreasing age at puberty in gilts, relative to biological and economic efficiency in pork production. Acccording to Kirkwood and Hughes (1982) many pig-breeding herds replace about 30% of their breeding females each year. It has been estimated in the UK that between 15% and 25% of litters produced in pig units are from gilts; because gilts have smaller litters and longer weaning to rebreeding intervals than sows, the percentage of gilts in a herd can have a marked effect on overall productivity. The efficiency of reproduction in the pig unit can be still further reduced if gilts are present in the herd for prolonged periods prior to first mating.

For some time, there has been a great deal of research interest in ways and means of reducing the age at first farrowing as a method of producing more pigs per breeding animal each year; as part of this, techniques for advancing and synchronizing the onset of puberty in gilts could be of practical interest.

9.1.1. Puberty and lifetime production

As observed by Cole (1975), the general rule in practice in the UK has been to mate the gilt for the first time at a liveweight between 120 and 135 kg and after one or two heat periods have passed. In approaching the question of the commercial merit of earlier breeding, it is clearly desirable to have ample information on factors likely to influence the attainment of puberty in the gilt. It is also important to examine lifetime reproductive performance in relation to age at first mating. In Taiwan, for example, Huang and Lee (1995) first mated gilts of three breeds (Landrace, Yorkshire and Duroc) at ages ranging from <250 days to >330 days of age; they found no adverse effects over four parities from the practice of early mating. On the basis of their evidence, they suggested that the optimal age at first mating could be less than 8 months. In the Netherlands, Schukken *et al.* (1994) concluded that under their herd conditions the optimal economic age at first conception was 200–220 days.

9.2. Physiology and Endocrinology of Puberty in Gilts

Although it can be readily demonstrated that the ovaries of the gilt are capable of responding to exogenous gonadotrophins well in advance of the time when the animal reaches natural puberty, it is also true that pigs differ in some respects from cattle and sheep in what can be done in obtaining embryos from the very young prepubertal animal.

9.2.1. Species differences

In contrast to heifer calves and ewe lambs, investigators have usually failed to induce follicular development or ovulation in piglets in the first 2 months or more of life after treatment with gonadotrophins (Casida, 1935; Oxender *et al.*, 1979); this would agree with the observation that vesicular (tertiary) follicles do not appear in the ovaries of the gilt until it is more than 60 days old. As noted elsewhere (Chapter 7), there is evidence that the pig differs from many other mammals in its pattern of oogenesis (Fulka *et al.*, 1972), which apparently lasts until some weeks after birth.

Gilts usually reach puberty between 6 and 8 months of age. Although considerable variation may exist in the time of heat onset according to genetic and environmental factors, the dominant changes that initiate puberty are believed to occur in the hypothalamus. It is clear that such changes are mediated by way of interactions between the animal's internal and external environment, the extrahypothalamic brain centres, the hypothalamic–pituitary axis and the ovaries.

Available evidence in the gilts suggests that the onset of puberty is neither limited by the functioning of the stimulatory oestrogen feedback mechanisms nor by the secretory activity of the pituitary or the ovary; each of these mechanisms can be activated by treatment with appropriate hormones. Studies by Foxcroft *et al.* (1984) led them to conclude that several characteristics of the positive feedback mechanism undergo maturational changes between 60 and 160 days of age in the domestic gilt; these changes are believed to be ovarian and particularly oestrogen dependent. Findings reported by Dial *et al.* (1984) indicated that the maturation of the pig ovary may limit the age at which puberty occurs. It would appear that in the pig, as in the other farm mammals, puberty is brought about by a reduction in the activity of intrinsic neural inhibitory mechanisms and/or a decrease in the negative feedback action of ovarian steroids; such events result in the stimulation of pulsatile GnRH release and consequent augmentation of episodic LH-secretion, which in turn leads to ovarian activity.

9.2.2. *Meishan as useful model*

According to Prunier *et al.* (1993), the Meishan breed is likely to be a useful model to investigate the endocrine mechanisms that control the onset of puberty; they recorded that hormonal changes in Meishan gilts showed a similar pattern to those observed in European breeds.

9.2.3. *Gonadostat hypothesis*

The 'gonadostat' hypothesis proposed by Ramirez and McCann (1963) has been widely accepted regarding the endocrine control of puberty in mammals. In the USA, Lutz *et al.* (1984) showed that the onset of puberty in the gilt was preceded by an increase in LH and oestradiol concentrations, suggesting that the gilt also conforms to the 'gonadostat' hypothesis; their evidence indicated that the prepubertal increase in LH concentration was mediated by an increased frequency of lower amplitude LH peaks.

9.2.4. *Oocyte quality at puberty*

It is well established that fertility in the young pig may be lower at the first oestrus than at subsequent heat periods. In Oregon, Koenig and Stormshak (1993) evaluated oocytes recovered from gilts at the pubertal and third oestrus; they found that the percentage of immature oocytes (between germinal vesicle and prometaphase stages) was significantly greater (22% vs. 12%) at the first than at the third oestrus; such results indicate that oocytes of first oestrus gilts are inferior to those of third oestrus animals. These findings would support evidence from previous studies, such as those of Graham *et al.* (1990), who found that embryos from first oestrus gilts exhibited decreased development *in vitro* compared to embryos from third oestrus gilts.

Not all reports have shown evidence of increased reproductive performance as gilts pass through heat periods after reaching puberty. In Northern Ireland, Walker *et al.* (1989) stimulated oestrus with mature boars and bred gilts at either the first or second heats; the number of piglets born in litter to first oestrus gilts was 9.7 and 10.4 to second oestrus animals.

9.2.5. *Predictors of puberty*

There were indications in some studies that gilts with LH levels above 1 ng ml^{-1} at 80 days of age were those likely to achieve puberty before 200 days of age; further work by Woods *et al.* (1995) failed to confirm this although they did record that vulva size at 155 days of age was indicative of early puberty. An earlier paper by Eliasson (1991) in Sweden had shown that gilts which attained puberty earlier showed more intense and longer duration of reddening and swelling of the vulva.

9.3. Nutritional, Environmental and Breed Factors

In wild pig communities, young gilts would normally reach puberty in the late autumn, at about 8 months of age; it is generally accepted that in such communities the attainment of puberty is more closely related to age than to liveweight. This may help to explain why, under normal commercial feeding regimes, the effect of diet on puberty may not be of such great importance.

9.3.1. *Effect of nutrition*

According to Dziuk (1977), the plane of nutrition used in rearing gilts may influence the age at which first oestrus is shown to a minor extent. Much of the variability of age at puberty appears to be related to the genetic background of the gilt and to other aspects of the environment in which she is kept. Friend *et al.* (1981, 1986) showed that gilts offered feed *ad libitum* can reach puberty at an earlier age than those restricted in feed intake. Low-plane feeding which markedly reduces growth rates in gilts has certainly been shown to delay puberty (Aherne and Kirkwood, 1985; Dyck, 1988).

Even relatively modest reductions in the feed intake of modern pig genotypes has been shown to delay puberty by more than 3 weeks (Van Lunen and Aherne, 1987). A review by King (1989b) recommended that replacement gilts intended for breeding should receive at least 36 MJ digestible energy per day to ensure that they reach 120–125 kg liveweight at 200 days of age. From about 2 weeks before mating, the gilts should be placed on a finisher diet (0.55 g of available lysine per MJ digestible energy) *ad libitum* to ensure optimum ovulation rates at mating.

The effect of body composition on the timing of puberty in pigs is still in some doubt. There have been reports of a positive relationship between backfat thickness and age of puberty in gilts fed on a high plane of nutrition but not for those reared on a low plane (Den Hartog and Noordewier, 1984). There have also been reports that high levels of body fat at a young age may delay puberty (King, 1989a). How far protein intake may influence the onset of puberty in pigs is still in some doubt. There have even been suggestions that supplementing the diet with tyrosine, a precursor of the brain neurotransmitters, dopamine and noradrenaline, may also advance the age of puberty in the gilt.

According to Robinson (1990), however, the range of feeding levels adopted during the rearing period in modern commercial pig-production systems is unlikely to alter puberty in the gilt by more than a few weeks.

9.3.2. Breed effects

Mean values of around 200–210 days for the age of pigs at natural puberty in different breeds of pig have been reported (reviews by Anderson and Melampy, 1972). Several papers from the UK and USA in the 1970s showed the normal range of ages at puberty as being from about day 140 to day 250, with around 200 days being the average value (Dyck, 1971; Cunningham *et al.*, 1974; Callaghan and King, 1977). Allrich *et al.* (1985) showed the average age of puberty to be significantly greater in Yorkshire gilts in comparison with Landrace animals (200 vs. 185 days) and it is generally accepted that crossbred gilts reach puberty earlier than purebred animals.

The most dramatic evidence on breed effect comes from studies with the Chinese Meishan which have shown that the Meishan gilt only requires about half the time needed for European-type gilts to reach puberty (Cheng, 1983). At Nottingham, Hunter *et al.* (1993) recorded an average age of puberty in Meishan gilts of 115 days in contrast to a value of 235 days in Large White crossbreds. Other studies at Nottingham recorded similar evidence of early puberty in the Meishan (Faillace *et al.*, 1994).

9.3.3. Puberty in total confinement systems

Delayed puberty has been shown to be a problem to pig producers using total confinement systems of management and it has been suggested in several reports that this may be the result of the environment imposing a physiological stress on the gilts (Jensen *et al.*, 1970; Rampacek *et al.*, 1981). Although there have been indications that group-penned gilts reach puberty a little earlier than tethered pigs, when Ford and Teague (1978) investigated the effects of crowding gilts in a total confinement system to the point that growth rate and feed efficiency was depressed, they did not find that the age of puberty was affected. In Iowa, on the other hand, Awotwi and Anderson (1985) reported that a higher percentage of group-reared gilts were cycling at 8 months than of those reared in isolation (75% vs. 50%).

9.3.4. Space restrictions

In Australia, Pearce and Paterson (1993) examined the effect of space restriction during rearing on the attainment of puberty and subsequent reproductive activity of pigs. Although crowded gilts reached puberty as early as uncrowded animals, mating rate and pregnancy rate tended to be lower and fewer crowded gilts exhibited a first oestrous cycle of normal length.

9.3.5. Seasonal and light effects

The extent to which season may influence puberty in pigs has been the subject of various reports. Gossett and Sorensen (1959) and Sorensen *et al.* (1961) found no difference in the age of puberty between spring- and autumn-born animals; in contrast, autumn-born pigs reached puberty earlier than those born in spring in a report by Mavrogenis and Robison (1970). Elsewhere, evidence has indicated that the proportion of gilts reaching puberty increased during periods of increasing daylength (Scanlon and Krishnamurthy, 1974). Possible breed–seasonal interactions were evident in data reported by Zimmerman *et al.* (1960), who recorded spring-born Chester White gilts reaching puberty earlier than autumn-born animals whereas Poland China pigs showed the reverse. Seasonal trends were also evident in data reported by Caton *et al.* (1986) in Missouri.

In Australia, Paterson *et al.* (1989) concluded that the threshold level of response by gilts to the 'boar effect' changed throughout the year, being highest in summer and lowest in spring. In Nottingham, Faillace *et al.* (1994), working with Meishan pigs, recorded that gilts approaching puberty in the spring were significantly younger and more likely to reach puberty by 100 days of age than those approaching puberty in the autumn.

9.3.6. Light manipulation

A number of studies have suggested that supplemental lighting may affect growth and subsequent reproductive performance in pigs. Although there is evidence that supplemental lighting may hasten the onset of puberty in the boar (Hoagland and Diekman, 1982), the results on the effect of photoperiod on the onset of puberty have been less clear. Dufour and Bernard (1968) in Canada reported that short daily photoperiods hastened the onset of first oestrus in gilts; on the other hand, others have reported that long photoperiods hastened puberty (Ntunde *et al.*, 1979), or that supplemental lighting was without effect (Wise *et al.*, 1980). In Australia, Paterson and Pearce (1990) examined the attainment of puberty in gilts reared under long-day or short-day artificial light regimens; their results suggested that, at cool temperatures, long days delayed the timing of puberty. Results presented by Awotwi and Anderson (1985), on the other hand, agreed with those of Wise *et al.* (1980) and Diekman and Hoagland (1983) in finding that photoperiod had no effect on the age at puberty in gilts.

9.3.7. Stress effects

It has been reported that various forms of stress can stimulate the onset of puberty in pigs. The stress resulting from mixing, relocating and transporting gilts has been demonstrated to stimulate earlier puberty (Du Mesnil du

Buisson and Signoret, 1962; Bourn *et al.*, 1974; Zimmerman *et al.*, 1976). There is also evidence suggesting that the combination of boar exposure plus the stress of mixing or transport may be more effective in stimulating earlier puberty in gilts than boar exposure alone (Bourn *et al.*, 1974; Eastham and Cole, 1983). In Cambridge, however, Stephens and Close (1984) found that neither real nor simulated transport of gilts had any apparent effect on age at puberty in gilts. Work in Leeds also showed that a change to a different pen or to a recently vacated boar pen alone were ineffective stimuli for inducing an earlier onset of puberty (Pearce and Hughes, 1985). It is clear that transport and relocation effects may be more complex than was once believed.

The effect of elevated environmental temperature on hormone profiles and the onset of puberty was examined by Flowers and Day (1990); gilts were kept at 15.6° or 33°C from 150 to 180 days of age. It was concluded from the study that delayed puberty in gilts kept at an elevated environmental temperature was partly due to decreased gonadotrophin secretion.

9.4. The Boar Effect

Work reported from the late 1960s onwards by many authors showed that the timing of puberty in gilts could be markedly influenced by the age at which they first had contact with the boar (Brooks and Cole, 1970; Bourn *et al.*, 1974; Mavrogenis and Robison, 1976; Kirkwood *et al.*, 1981, 1983; Cronin, 1983; Pearce and Hughes, 1985; Caton *et al.*, 1986; Paterson *et al.*, 1991).

9.4.1. *Applying the boar effect*

The essential steps in this, under Irish and UK conditions, include keeping the gilt out of sight, sound and smell of the boar until she reaches about 160 days of age; she then gets a change of environment and is exposed to a mature sexually active boar, an experience which can have the effect of bringing her and companion females in oestrus within 1 week. The exposure to the boar may take the form of 30 minutes each day or it may be continuous. Although young boars can induce a response, they may be less effective than an older, mature boar (Fig. 9.1).

The boar effect can reduce the age at first service (at the second or third oestrus) by several weeks; it can also mean a smaller pig producing the first litter and lower feed costs for the replacement gilt. In Australia, Cronin (1983) attempted to quantify the boar effect by showing that exposure of gilts to mature boars at 160 rather than 200 days increased the reproductive efficiency of the pigs to the extent of an additional 0.24 piglets per gilt.

Exposure methods

Results reported by Van Lunen and Aherne (1987) in Canada indicated that continuous boar exposure for a 10-day period or a 30-minute daily

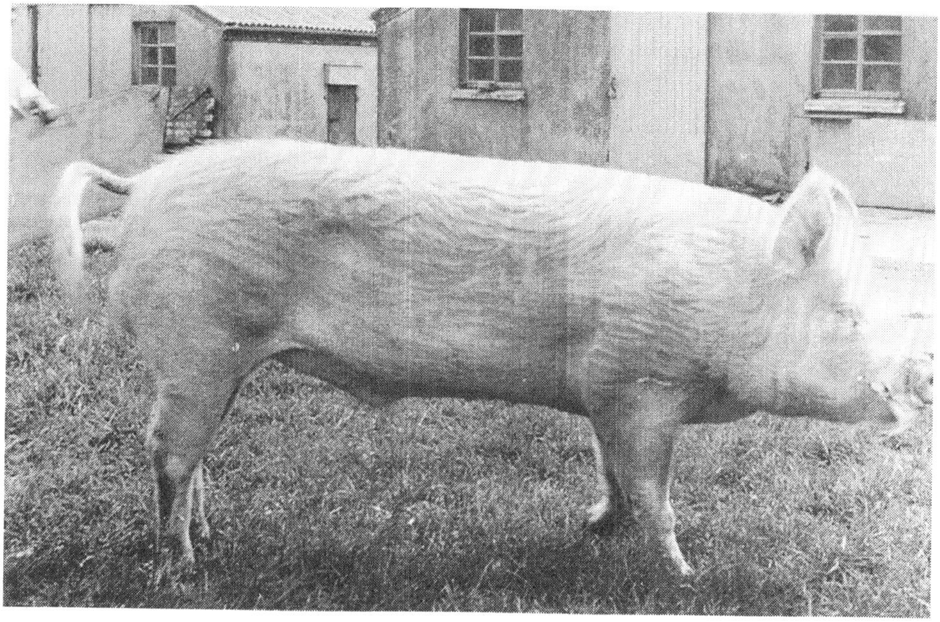

Fig. 9.1. The 'boar effect' requires a male that is suitably mature if it is to produce a response in the gilt.

boar exposure for 10 days by moving the gilts to the boar induced a greater percentage of gilts to reach earlier puberty in comparison with other boar exposure treatments evaluated. In Australia, Paterson *et al.* (1989) showed that when exposure to the boar was limited to 1 day or 10 consecutive days, although a degree of stimulation was achieved, daily exposure until puberty occurred was necessary for the maximum stimulation of puberty. In Germany, Mosch and Huhn (1992), working with gilts aged 159–180 days, found that direct physical contact with boars and continuous or limited exposure over a week in a 3-week cycle were more effective in inducing puberty than was less intensive contact such as walking the boars past the gilt's pen.

Effect of libido

The influence of libido on the efficacy of the boar effect was clearly demonstrated in studies reported by Hughes (1994) in Australia in which exposure to a high-libido boar stimulated puberty at a significantly earlier age than exposure to a low-libido boar (180 vs. 194 days).

In Australia, Philip and Hughes (1995) have reported on the effects of contact frequency and season on the efficacy of the boar in inducing earlier puberty in gilts exposed at an average age of 160 days. They found that the proportion of gilts reaching puberty within the 60 day experimental period increased as the frequency of boar contact increased; there was no significant effect of season on the timing of puberty.

9.4.2. Rearing of gilts

In some of the early studies at Nottingham, there were indications that introducing the boar to gilts which were too young might delay puberty due to habituation to the presence of the male. In studies which examined this question, gilts were either reared with castrate male contemporaries or in total isolation from male pigs (Aherne *et al.*, 1976) or were reared with intact male pigs (Walker, 1979; Paterson and Lindsay, 1980). The gilts responded to the introduction of the mature boar just the same whether they had been reared with castrate or intact male companions. In Nottingham, Cole *et al.* (1982) and Eastham *et al.* (1984) also reported that gilts reared with contemporary males reached puberty round the same time as gilts reared alone. For the pig producer, it was useful to know that the practice of not castrating male pigs reared for market is compatible with using the 'boar effect' to advance puberty in gilts.

Later studies at Nottingham by Eastham and Cole (1987) showed that rearing gilts in fence-line contact with a mature boar did not appear to have any deleterious effects on the attainment of puberty when the puberty stimulus applied was a combination of relocation and full physical contact with a mature boar (either a novel boar or the same boar that was present during rearing).

9.4.3. Source of stimulus

Kirkwood *et al.* (1981) indicated a significant role for olfaction in the precocious induction of puberty in gilts kept in close proximity to a boar. Their study showed that the 'boar effect' could be completely eliminated by rendering prepubertal gilts anosmic. The way in which the boar achieves his effect appears to differ from that recorded in mice, where the source of primer pheromone is known to be in the male's urine. In the boar, however, studies utilizing either isolated urinary pheromones (known to have signaller pheromone properties) or boar urine, have not been successful in advancing puberty.

Although Kirkwood and Hughes (1982), in a review of the role of the boar in puberty in the gilt, initially suggested that a primer pheromone responsible for the 'boar effect' was released from the submaxillary gland of the boar, later work by Booth (1984) and Kirkwood *et al.* (1983) made it appear unlikely that the signalling pheromones 3-α-androstenol and 5-α-androstenone in boar saliva or other constituents of the saliva play a major role as primer pheromones in the 'boar effect'. In Aberdeen, Deligeorgis *et al.* (1984a,b) compared the efficacy of the 'boar effect' when the tactile component of the stimulus was absent, except that which could be applied through a wire-mesh fence. They found that full contact with a mature boar was more effective in inducing precocious puberty than only partial contact, whatever the nature of the stimuli might be.

Booth (1984) concluded that if olfactory cues do play a major role in the male-effect, then they probably arise from sources other than saliva or urine. It was suggested that the olfactory component of the male-effect probably involves a complex of pheromones arising from a number of different sources in the boar. It was concluded that pheromones may have to act in concert with other, as yet undefined, sensory inputs (auditory, visual, tactile) to produce the full 'boar effect'. Studies later reported by Pearce and Hughes (1987) were to show that the 'boar effect' is mediated by several boar stimuli acting together. Tactile cues originating from the physical contact with the gilts are apparently necessary for olfactory cues to exert priming pheromone effects in the gilt. The olfactory stimulus is believed to be a combination of, as yet, undefined compounds in boar urine and the 16-androstenes, 3-α-androstenol and 5-α-androstenone present in the boar submaxillary gland saliva.

Importance of physical contact

According to Hughes *et al.* (1990), in a comprehensive review of mechanisms mediating the stimulatory effect of the boar on gilt reproduction, knowledge has expanded over the years to include various factors which interact with the 'boar effect' to influence the efficacy of the male stimulus. Stimuli emanating from the boar are in two categories, olfactory and tactile/stress cues. The same authors note that the endocrine basis of the 'boar effect' remains obscure. In a study that examined the endocrine and pubertal response of gilts to boar exposure, Pearce and Paterson (1992) showed that physical contact with the boar was essential because it allowed the transfer of boar pheromones.

9.4.4. Exposure of gilts to large-stimulus pigs

The effect of daily exposure to oestrous or anoestrous pigs on the attainment of puberty in the gilt was the subject of a report by Peacock and Hughes (1995). In this, gilts (159 days old) exposed to large (192 kg) stimulus females reached puberty at an earlier age (204 vs. 216 days) than those exposed to small (78.9 kg) stimulus females; the sexual status of the exposure female had no effect on the timing of puberty.

9.4.5. Long-term effects

The long-term reproductive performance of gilts which have had puberty advanced by the 'boar effect' has been examined in several studies. These have shown that although earlier mating may result in a reduction in the size of the first litter, it is likely to result in improved productivity when other factors are taken into account. Brooks and Smith (1980) recorded that over five litters, the earlier mated gilts produced just as many piglets and showed a 6% better food conversion efficiency than controls. However, they also noted that producers contemplating earlier breeding of gilts should not cull on the basis of first litter performance as they might otherwise do with normal-aged animals. In Ireland, Lynch *et al.* (1994) examined the effect of age at first mating on lifetime performance of sows. In this, gilts were isolated from male pigs from 30 kg liveweight until boar exposure at 200–240 days of age; results suggested there was no long-term benefit in delaying mating of maiden gilts beyond 200 days of age.

9.4.6. *Type of housing*

An experiment by Prunier and Meunier-Salaun (1989) sought to determine the influence of boar exposure on the age of puberty in tethered and group-penned gilts. They showed that the effect of boar introduction was more important in stimulating earlier puberty than the type of housing used. They did record an interaction between 'boar effect' and type of housing; they suggested that in tethered gilts there might be a risk of boar introduction becoming ineffective when other negative factors are present (e.g. altered behavioural responses).

9.4.7. *Other considerations*

Before leaving the question of natural puberty in pigs, it should be mentioned that some proportion of gilts may not show their first oestrus until as late as 9 months and for that reason may be considered abnormal. Swedish workers in the 1970s estimated that about 12% of gilts in breeding herds in that country fell into that category.

9.5. Hormonal Induction of Puberty and Pregnancy in Gilts

After the age of about 3 months, it is possible to induce follicle development, oestrus and ovulation in the prepubertal gilt with gonadotrophins such as PMSG, as demonstrated first more than 60 years ago (Casida, 1935); if such animals are bred, fertilization and early embryonic development does occur. However, whether the gilt remains pregnant or not may largely be a question of its physiological age at the time of treatment. In this regard, the pig is apparently unlike the prepubertal sheep, in which it is possible to establish pregnancy and see it maintained in the young ewe without any great difficulty.

9.5.1. Hormonal regimens

Early treatments with gonadotrophins often took the form of PMSG followed by hCG or a combined PMSG/hCG treatment. Dziuk and Gehlbach (1966) reported that 90% of gilts (90–130 days old; 45–55 kg liveweight) were induced to ovulate by treatment with PMSG followed by hCG; however, the proportion of gilts showing oestrus at the time of breeding or remaining pregnant afterwards was low; other workers at that time reported much the same. It has been suggested that the problem in the gilts might be due to the uterus being insufficiently developed to allow normal implantation and to prevent the process of luteolysis (McMenamin and King, 1974); however, treatments applied in an effort to promote uterine development apparently failed to achieve any improvement. It has also been suggested that the corpora lutea of the prepubertal gilt may not be maintained by pituitary luteotrophins unless the animal is close to normal puberty (Segal and Baker, 1973).

In Missouri, Flowers *et al.* (1989) drew attention to the fact that, despite its widespread use in prepubertal, cyclic and anoestrous gilts and sows, the manner in which PMSG stimulates ovarian activity in the pig is not fully understood. In a study to determine whether some part of the PMSG-induced follicular development in pigs was due to a stimulation of FSH secretion, they showed that it was unlikely that any stimulation of FSH is involved in the ability of PMSG to induce follicular growth and ovulation.

9.5.2. PMSG/hCG combination

For reasons such as those already described, the use of gonadotrophins in prepubertal gilts must be approached with due caution. On the other hand, there have been reports of acceptable results with a PMSG/hCG combination. Schilling and Cerne (1972) used such a combination (400 IU PMSG + 200 IU hCG) in gilts aged 150–165 days old. The animals were bred by AI on the fourth and fifth days afterwards and 66% of the females were pregnant at slaughter a month after breeding. On the other hand, low pregnancy rates, of the order of 10–33%, were reported by Rampacek *et al.* (1976) in similarly aged gilts (155–175 days) that received this same form of treatment.

The PMSG/hCG combination was used in the 1970s with varying degrees of success by Canadian workers (Baker and Downey, 1975) and reported on by several Eastern European authors (Miskovic and Stancic, 1978; Schlegel *et al.*, 1979); the general finding was that pregnancy rate and litter size from the combination treatment improved the closer the gilt approached the natural age of puberty. A report by Glei *et al.* (1990) in the former East Germany emphasized that oestrus should not be induced in gilts weighing less than 100 kg. Huhn *et al.* (1994), also dealing with work in the former East Germany, concluded that the routine induction of oestrus had no adverse effect on subsequent reproductive performance.

In Ireland, studies on the induction of oestrus in gilts (70–90 kg body weight), using either 750 IU PMSG or PG600 with 500 IU hCG given after a further 72 h showed that of the gilts treated with PMSG and PG600, 20/20 (100%) and 16/20 (80%), respectively, showed oestrus (Stephens *et al.*, 1988).

9.5.3. Once-bred gilt systems

Accelerated puberty is clearly relevant to the feasibility of the once-bred gilt system, where the idea is to take a litter from the pig before dispatching her to the heavy hog market. Greater flexibility of curing procedures also enables many more options for large joints and cuts. Stewart *et al.* (1993) drew attention to the fact that moves towards high welfare systems as an alternative to sow stalls for pregnant female pigs make the system of once-bred gilts very attractive. In this, however, it is important to have a predictable throughput of relatively young gilts which are mated on at least their second observed heat. The Aberdeen workers investigating the potential of using PG600 to synchronize prepubertal gilts in group-housed feeding systems found that 87% of gilts showed heat within a week and that the synchrony was maintained through to service at the second oestrus in most animals. Dietary effects on oestrus expression and ovulation rate in PG600-treated gilts were subsequently examined by workers at Aberdeen (Glasgow *et al.*, 1996); diets containing a high proportion of fibre did not affect ovarian activity in the animals.

9.5.4. hCG alone

Mattioli *et al.* (1991) in Italy explored the possiblity of using hCG alone, rather than in combination with PMSG, to induce follicular growth and ovulation in prepubertal gilts. This was done on the understanding that there is much evidence to show that the limiting factor to ovarian activity is due to LH pulse frequency and amplitude rather than to a lack of FSH action; in a small-scale experiment, it was found that hCG alone was sufficient to induce oestrus and ovulation.

9.5.5. Maintenance of cyclical activity

Various studies looked not only at the question of oestrous response immediately after gonadotrophin treatment, but at the maintenance of regular cyclical activity in the gilts (Fig. 9.2). Paterson and Lindsay (1980) showed that when prepubertal gilts were housed in contact with mature boars, there was a much greater tendency for cyclical breeding activity to be maintained than when animals were kept isolated from the males (85% vs. 45%). They found that the ovulation rate varied widely at the first (induced) heat but at the second was much more consistent and normal. Such results were taken to

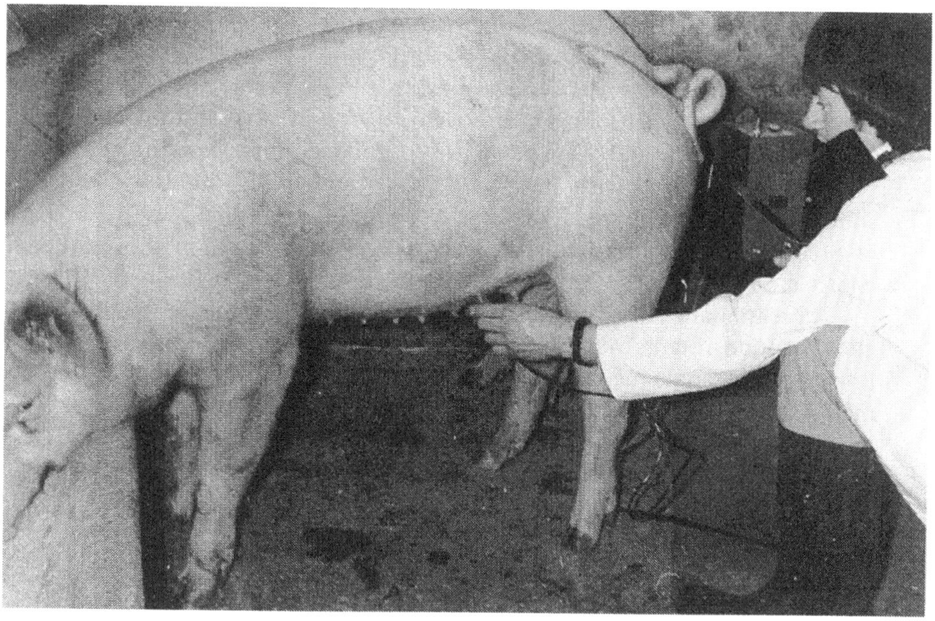

Fig. 9.2. Maintenance of pregnancy may be a problem in the gilt induced to show oestrus.

indicate that the PMSG/hCG combination could be of commercial value in inducing early puberty if cycles were maintained and the gilts mated at the second occurring heats.

In Northern Ireland, Burnett (1988) demonstrated that the ability of prepubertal gilts to respond to PG600 and to maintain cyclic activity to a second ovulation improved with increasing age and the stage of growth at treatment; the maintenance of cyclic activity was improved by housing gilts in fence-line contact with mature boars.

9.5.6. Gonadotrophins and supplemental progesterone

The enhancement of pubertal fertility with supplemental progesterone treatment was reported by Pope *et al.* (1993) in Ohio. Allowing a 3-day interval between injection with progesterone and PG600 (400 IU PMSG + 200 IU hCG) resulted in a higher fertility rate than a 1- or 2-day interval between injections. Progesterone priming prior to the administration of PMSG/hCG was also found (Rajamahendran *et al.*, 1992) to be the most effective of various treatments for the maintenance of cyclic activity.

9.5.7. Gonadotrophin in combination with prostaglandin

There has been some evidence to suggest that $PGF_{2\alpha}$ or an analogue could be used to advance, synchronize or induce ovulation in gilts. A study in Canada by Srikandakumar and Downey (1989), for example, working with pre-pubertal gilts treated with PMSG (750 IU) and hCG (500 IU after 72 h) showed that when 500 μg cloprostenol was injected 34–36 h after hCG, 78% of preovulatory follicles ovulated in comparison with 0% in the untreated controls.

A preliminary report on a new method for the synchronization of oestrus and ovulation in gilts was made by Schnurrbusch *et al.* (1994), who used a prostaglandin analogue (tiaprost) in combination with PMSG/hCG. Treatment consisted of induction of puberty in prepubertal gilts by simultaneous injection of PMSG/hCG at the age of 180–200 days. Regression of the first crop of corpora lutea was induced by PG in the late luteal phase and follicular growth was again stimulated with PMSG and ovulation controlled by hCG. All the gilts showed a synchronous oestrus in the second cycle; ovulations were predicted to occur about 38–42 h after injection of the hCG.

9.5.8. Oestrogen treatment

As well as the gonadotrophin approach, other studies reported the use of steroid hormones as a means of activating the hypothalamic–pituitary axis. Some of the earliest work was in Canada, where the steroid 5-α-androstane-3 β-17 β-diol was administered without success to gilts varying in age from 109 to 180 days (Callaghan and King, 1977). In the UK, treatment of gilts generally involved six doses of oestradiol benzoate (0.2 mg per dose) given over 3 days (Hughes and Cole, 1978; Paterson and Day, 1980; Paterson *et al.*, 1984). Although treatment in the original work at Nottingaham was effective (60% responding within 5 days) in 140-day-old gilts (Hughes and Cole, 1978), such findings were not substantiated in work elsewhere.

It was evident that the proportion of gilts ovulating after oestradiol benzoate treatment could be extremely variable; some suggested that this variability might be attributable to differences in the stage of sexual maturity at the time of treatment (Deligeorgis *et al.*, 1984). However, Yang *et al.* (1988), at Leeds, showed that the proportion of prepubertal gilts responding to oestradiol treatment did not increase as the age of gilts increased. The same workers examined the effect of five different doses of oestrogen but found no indication that dose level influenced the subsequent reproductive performance of gilts in terms of ovulation rate and embryo survival (Yang *et al.*, 1987). Some of their data supported the view that ovulation rate is linked to the number of ovulatory cycles rather than to the age or weight of the gilt. The general conclusion with oestrogen treatment is that results are too unpredictable for it to be of commercial interest.

9.5.9. Melatonin treatment

The ability of melatonin, applied orally or by implant, to influence puberty in gilts, has been investigated in various countries. In the USA, Diekman *et al.* (1991), in a report on a reduction in the age of puberty in gilts consuming melatonin (3 mg day^{-1}) during periods of increasing or decreasing daylength, showed that age of puberty was significantly lower in melatonin-treated animals than in controls (198 vs. 228 days). It was concluded that the daily administration of 3 mg melatonin at 15.30 h advanced the onset of puberty in gilts, regardless of daylength. In Australia, Paterson *et al.* (1992) found that, in contrast to constant-release melatonin implants, oral melatonin given each afternoon prevented seasonal inhibition of puberty. These results also provided evidence that photoperiod mediates the seasonal onset of puberty by changes in the diurnal rhythm of plasma melatonin.

9.6. References

Aherne, F.X. and Kirkwood, R.N. (1985) Nutrition and sow prolificacy. *Journal of Reproduction and Fertility* 33, 169–183.

Aherne, F.X., Christopherson, R.J., Thompson, J.R. and Hardin, R.T. (1976) Factors affecting the onset of puberty, post-weaning estrus and blood hormone levels in Lacombe gilts. *Canadian Journal of Animal Science* 56, 681–692.

Allrich, R.D., Christenson, R.K. and Ford, J.J. (1985) Age at puberty and estrous activity of straightbred and reciprocal crossbred gilts. *Animal Reproduction Science* 8, 281–286.

Anderson, L.L. and Melampy, R.M. (1972) Factors affecting ovulation rate in the pig. In: Cole, D.J.A. (ed.) *Pig Production*. Butterworths, London, pp. 329–366.

Awotwi, E.K. and Anderson, L.L. (1985) Effect of lighting regimen, rearing in isolation and olfactory bulbectomy on growth and puberty in Yorkshire gilts. *Animal Reproduction Science* 9, 51–61.

Baker, R.D. and Downey, B.R. (1975) Induction of estrus, ovulation and fertility in prepubertal gilts. *Annales de Biologie animale Biochimie Biophysique* 15, 375–382.

Booth, W.D. (1984) A note on the significance of boar salivary pheromones to the male-effect on puberty attainment in gilts. *Animal Production* 39, 149–152.

Bourn, P., Carlson, R., Lantze, B. and Zimmermann, D.R. (1974) Age at puberty as influenced by age at boar exposure and transport. *Journal of Animal Science* 39, 987.

Brooks, P.H. and Cole, D.J.A. (1970) The effect of boar presence on the age at puberty of gilts. *Journal of Reproduction and Fertility* 23, 435–440.

Brooks, P.H. and Smith, D.A. (1980) The effect of mating age on the reproductive performance, food utilization and liveweight change of the female pig. *Livestock Production Science* 7, 67–78.

Burnett, P.J. (1988) A study of the effect of age and growth traits on the attainment of puberty and subsequent reproductive performance in gilts given boar stimulation with or without PMSG + HCG. *Dissertation Abstracts International B (Sciences & Engineering)* 49, 2005.

Callaghan, B. and King, G.J. (1977) Effect of 5-alpha-androstane-3 beta, 17 beta-diol on inducing early puberty in gilts. *Canadian Journal of Animal Science* 57, 599–600.

Casida, L.E. (1935) Prepubertal development of the pig ovary and its relation to stimulation with gonadotrophic hormone. *Anatomical Record* 61, 389–396.

Caton, J.S., Jesse, G.W., Day, B.N. and Ellersieck, M.R. (1986) The effect of duration of boar exposure on the frequency of gilts reaching first oestrus. *Journal of Animal Science* 62, 1210–1214.

Cheng, P-L. (1983) A highly prolific pig breed of China – the Taihu pig. *Pig News and Information* 4, 407–425.

Cole, D.J.A. (1975) Towards greater sow productivity. *Span* 18, 111–142.

Cole, D.J.A., Dyck, G.W. and Nathan, S. (1982) Reproduction in the gilt. 4. The effect of rearing gilts with contemporary castrated males or intact males on the attainment of puberty. *Animal Production* 35, 171–175.

Cronin, G.M. (1983) The effect of early contact with mature boars on reproductive efficiency in the gilt. *Animal Reproduction Science* 51–57.

Cunningham, P.J., Naber, C.B., Zimmerman, D.R. and Peo, E.R. (1974) Influence of nutritional regime on age at puberty in gilts. *Journal of Animal Science* 39, 64–67.

Deligeorgis, S.G., Lunney, D.C. and English, P.R. (1984a) A note of the efficacy of complete v. partial boar exposure on puberty attainment in the gilt. *Animal Production* 39, 145–147.

Deligeorgis, S.G., English, P.R., Lodge, G.A. and Foxcroft, G.R. (1984b) Comparison of two methods for evaluating reproductive development in prepubertal gilts. *Animal Production* 38, 283–291.

den Hartog, L.A. and Noordewier, G.J. (1984) The effect of energy intake on age at puberty in gilts. *Netherlands Journal of Agricultural Science* 32, 263–280.

Dial, G.D., Dial, O.K., Wilkinson, R.S. and Dziuk, P.J. (1984) Endocrine and ovulatory responses of the gilt to exogenous gonadotropins and estradiol during sexual maturation. *Biology of Reproduction* 30, 289–299.

Diekman, M.A. and Hoagland, T.A. (1983) Influence of supplemental lighting during periods of increasing or decreasing daylength on onset of puberty in gilts. *Journal of Animal Science* 57, 1235–1242.

Diekman, M.A., Clapper, J.A., Green, M.L. and Stouffer, D.K. (1991) Reduction in age of puberty in gilts consuming melatonin during decreasing or increasing daylength. *Journal of Animal Science* 69, 2524–2531.

Dufour, J. and Bernard, C. (1968) Effect of light on the development of market pigs and breeding gilts. *Canadian Journal of Animal Science* 48, 425–430.

Du Mesnil du Buisson, F. and Signoret, J.P. (1962) Influence de facteurs externes sur le dechechement de la puberte chez la truie. *Annales de Zootechnie* 11, 53–59.

Dyck, G.W. (1971) Puberty, post-weaning estrous and estrous cycle length in Yorkshire and Lacombe swine. *Canadian Journal Animal Science* 51, 135–140.

Dyck, G.W. (1988) Factors influencing sexual maturation, puberty and reproductive efficiency in the gilt. *Canadian Journal of Animal Science* 68, 1–13.

Dziuk, P.J. (1977) Reproduction in pigs. In: Cole, H.H. and Cupps, P.T. (eds) *Reproduction in Domestic Animals*, 3rd edn. Academic Press, New York, pp. 456–474.

Dziuk, P.J. and Gehlbach, G.D. (1966) Induction of ovulation and fertilization in the immature gilt. *Journal of Animal Science* 25, 410–413.

Eastham, P.R. and Cole, D.J.A. (1983) The influence of boar presence and transport on the attainment of puberty in gilts. *Animal Production* 36, 519.

Eastham, P.R., Dyck, G.W. and Cole, D.J.A. (1984) Reproduction in the gilt. 5. The effect of the presence of the mature male during rearing on puberty attainment. *Animal Production* 39, 277–282.

Eastham, P.R. and Cole, D.J.A. (1987) Reproduction in the gilt. 7. Exposure of young gilts to the same mature boar and relocation: effects on the attainment of puberty. *Animal Production* 44, 435–441.

Eliasson, L. (1991) Relationship between puberty and production traits in the gilt. 2. Oestrous symptoms at puberty. *Animal Reproduction Science* 25, 255–264.

Faillace, L.S., Biggs, C. and Hunter, M.G. (1994) Factors affecting the age at onset of puberty, ovulation rate and time of ovulation in Chinese Meishan gilts. *Journal of Reproduction and Fertility* 100, 353–357.

Flowers, B., Martin, M.J., Cantley, T.C. and Day, B.N. (1989) The effect of pregnant mare serum gonadotropin on follicle stimulating hormone and estradiol secretion in the prepubertal gilt. *Animal Reproduction Science* 21, 93–100.

Flowers, B. and Day, B.N. (1990) Alterations in gonadotropin secretion and ovarian function in prepubertal gilts by elevated environmental temperature. *Biology of Reproduction* 42, 465–471.

Ford, J.J. and Teague, H.S. (1978) Effect of floor space restriction on age at puberty in gilts and on performance of barrows and gilts. *Journal of Animal Science* 47, 828–832.

Foxcroft, G.R., Elsesser, F., Stickney, K., Haynes, N.B. and Back, H.L. (1984) Ovarian oestrogen-dependent maturation of the LH/FSH surge mechanism during prepubertal development in the gilt. *Journal of Endocrinology* 101, 371–380.

Friend, D.W., Lodge, G.A. and Elliot, J.I. (1981) Effects of energy and dry matter intake on age, body weight and backfat at puberty and on embryo mortality in gilts. *Journal of Animal Science* 53, 118–124.

Friend, D.W., Wolynetz, M.S. and Robertson, H.A. (1986) Effect of feeding frequency on age and weight of confied gilts at puberty, and some related breeding phenomena. *Animal Reproduction Science* 11, 69–74.

Fulka, J., Kopecny, V. and Trebichavsky, J. (1972) Studies on oogenesis in the early post-natal pig ovary. *Biology of Reproduction* 6, 46–50.

Glasgow, V.L., Stewart, A.H. and Edwards, S.A. (1996) The effects of dietary fibre and puberty induction method on ovarian activity of gilts. *Proceedings of the British Society of Animal Science* (Winter Meeting), Paper 145.

Glei, M., Weissenborn, B. and Etzrodt, F. (1990) Increasing efficiency by means of partial induction of puberty in gilts. *Tierzucht* 44, 249–251.

Gossett, J.W. and Sorensen, A.N.Jr (1959) The effects of two levels of energy and seasons on reproductive phenomena of gilts. *Journal of Animal Science* 18, 40–47.

Graham, J.R., Ray, L.J. and Gardiner, C.S. (1990) Effect of maternal estrus and nutrient intake on in vitro development of pig embryos. *Journal of Animal Science* 68 (Suppl. 1), 431.

Hoagland, T.A. and Diekman, M.A. (1982) Influence of supplemental lighting during increasing daylength on libido and reproductive hormones in prepubertal boars on growth and puberty in boars. *Journal of Animal Science* 55, 1483–1489.

Huang, Y.H. and Lee, K.H. (1995) Effects of age of conception at first litter on post-reproductive performance of purebred gilts. *Journal of the Chinese Society of Animal Science* 24, 31–49.

Hughes, P.E. (1994) The influence of boar libido on the efficacy of the boar effect. *Animal Reproduction Science* 35, 111–118.

Hughes, P.E. and Cole, D.J.A. (1978) Reproduction in the gilt. 3. The effect of exogenous oestrogen on the attainment of puberty and subsequent reproductive performance. *Animal Production* 27, 11–20.

Hughes, P.E., Pearce, G.P. and Paterson, A.M. (1990) Mechanisms mediating the stimulatory effects of the boar on gilt reproduction. *Journal of Reproduction and Fertility* Suppl. 40, 323–341.

Huhn, U., Spitschak, K., Konig, I. and Mosch, W. (1994) Investigations on the age at puberty before and after long-term use of hormonal synchronization of oestrus in a herd of German Landrace sows. *Archiv fur Tierzucht* 37, 45–53.

Hunter, M.G., Biggs, C., Foxcroft, G.R., McNeilly, A.S. and Tilton, J.E. (1993) Comparisons of endocrinology and behavioural events during the periovulatory period in Meishan and Large-White hybrid gilts. *Journal of Reproduction and Fertility* 97, 475–480.

Jensen, A.H., Yen, J.T., Gehring, M.M., Baker, D.H., Becker, D.E. and Harmon, B.G. (1970) Effects of space restriction and management on pre and post-pubertal response of female swine. *Journal of Animal Science* 31, 745–750.

King, R.H. (1989a) Effect of live weight and body composition of gilts at 24 weeks of age on subsequent reproductive efficiency. *Animal Production* 49, 109–115.

King, R.H. (1989b) Nutritional management to improve the reproductive performance of commercial gilts. In: Barnett, J.L. and Hennessy, D.P. (eds) *Proceedings of the Biennual Conference of the Australasian Pig Science Association* (Albury, NSW), pp. 315–318.

Kirkwood, R.N. and Hughes, P.E. (1982) Puberty in the gilt – the role of boar stimulation. *Pig News and Information* 3(4), 389–394.

Kirkwood, R.N., Forbes, J.M. and Hughes, P.E. (1981) Influence of boar contact on attainment of puberty in gilts after removal of the olfactory bulbs. *Journal of Reproduction and Fertility* 61, 193–196.

Kirkwood, R.N., Hughes, P.E. and Booth, W.D. (1983) The influence of boar-related odours on puberty attainment in gilts. *Animal Production* 36, 131–136.

Koenig, J.L.F. and Stormshak, F. (1993) Cytogenetic evaluation of ova from pubertal and third-estrous gilts. *Biology of Reproduction* 49, 1158–1162.

Lunen, T.A.Van and Aherne, F.X. (1987) Influence of method of boar exposure on age at puberty of gilts. *Canadian Journal of Animal Science* 67, 553–556.

Lutz, J.B., Rampacek, G.B., Kraeling, R.R. and Pinkert, C.A. (1984) Serum luteinizing hormone and estrogen profiles before puberty in the gilt. *Journal of Animal Science* 58, 686–691.

Lynch, P.B., Hanrahan, T.J. and Arkins, S. (1994) Effect of age at first mating on lifetime performance of sows under two weaning regimes. *Proceedings of the British Society Animal Production* (Winter Meeting), Paper no. 170.

Mattioli, M., Galeati, G., Cappannari, C. and Parmeggiani, A. (1991) HCG induced follicular growth and ovulation in prepubertal gilts. *Journal of Reproduction and Fertility* Abstract Series no. 7, 58.

Mavrogenis, A.P. and Robison, O.W. (1976) Factors affecting puberty in swine. *Journal of Animal Science* 42, 1251–1255.

McMenamin, H. and King, G.J. (1974) Uterine development in prepubertal gilts after single or multiple stimulation by gonadotrophins. *Journal of Reproduction and Fertility* 39, 109–110.

Miskovic, M. and Stancic, B. (1978) Oestrus, ovulation and fertility in prepubertal gilts treated with exogenous gonadotrophins. *Veterinaria* 27, 69–76.

Mosch, W. and Huhn, U. (1992) The effect of boar contact and liveweight on puberty in gilts. *Archiv fur Tierzucht* 35, 385–398.

Ntunde, B.N., Hacker, R.R. and King, G.J. (1979) Influence of photoperiod on growth, puberty and plasma LH levels in gilts. *Journal of Animal Science* 48, 1401–1406.

Oxender, W.D., Colenbrander, B., Van de Wiel, D.F.M. and Wensing, C.J.G. (1979) Ovarian development in fetal and prepubertal pigs. *Biology of Reproduction* 21, 715–721.

Paterson, A.M. and Day, B.N. (1980) The effect of estradiol benzoate on estrus and ovulation in prepubertal gilts. *Journal of Animal Science* 51 (Suppl. 1), 88.

Paterson, A.M. and Lindsay, D.R. (1980) Induction of puberty in gilts. 1. The effects of rearing conditions on reproductive performance and response to mature boars after early puberty. *Animal Production* 31, 291–297.

Paterson, A.M. and Pearce, G.P. (1990) Attainment of puberty in domestic gilts reared under long-day or short-day artificial light regimens. *Animal Reproduction Science* 23, 135–144.

Paterson, A.M., Pearce, G.P., Foxcroft, G.R. and Reed, H.C.B. (1984) Reproductive performance of gilts induced into puberty with oestradiol benzoate or a combination of pregnant mare's serum gonadotrophin and human chorionic gonadotrophin. *Animal Production* 38, 121–128.

Paterson, A.M., Hughes, P.E. and Pearce, G.P. (1989a) The effect of season, frequency and duration of contact with boars on the attainment of puberty in gilts. *Animal Reproduction Science* 21, 115–124.

Paterson, A.M., Hughes, P.E. and Pearce, G.P. (1989b) The effect of limiting the number of days of contact with boars, season and herd of origin on the attainment of puberty in gilts. *Animal Reproduction Science* 18, 293–301.

Paterson, A.M., Pearce, G.P. and D'Antuono, M.F. (1991) Seasonal variation in attainment of puberty in isolated and boar-exposed domestic gilts. *Animal Reproduction Science* 24, 323–333.

Paterson, A.M., Maxwell, C.A. and Foldes, A. (1992) Seasonal inhibition of puberty in domestic gilts is overcome by melatonin adminsistred orally, but not by implant. *Journal of Reproduction and Fertility* 94, 97–105.

Peacock, A.J. and Hughes, P.E. (1995) The effects of daily exposure to oestrous or anoestrous gilts and sows on the attainment of puberty in the gilt. *Animal Reproduction Science* 40, 135–142.

Pearce, G.P. and Hughes, P.E. (1985) The influence of daily movement of gilts and the environment in which boar exposure occurs on the efficacy of boar-induced precocious puberty in the gilt. *Animal Production* 40, 161–167.

Pearce, G.P. and Hughes, P.E. (1987) The influence of boar-component stimuli on puberty attainment in the gilt. *Animal Production* 44, 293–302.

Pearce, G.P. and Paterson, A.M. (1992) Physical contact with the boar is required for maximum stimulation of puberty in the gilt because it allows transfer of boar pheromones and not because it induces cortisol release. *Animal Reproduction Science* 27, 209–224.

Pearce, G.P. and Paterson, A.M. (1993) The effect of space restriction during rearing on the attainment of puberty and subsequent reproductive activity of female pigs. *Animal Reproduction Science* 32, 99–106.

Philip, G. and Hughes, P.E. (1995) The effects of contact frequency and season on the efficacy of the boar effect. *Animal Reproduction Science* 40, 143–150.

Pope, W.J., Nephew, K.P. and Cardenso Seijas, H. (1993) Enhancement of pubertal fertility with supplemental progesterone treatment. *Ohio Swine Research and Industry Report* (1992–93), 85–87.

Prunier, A. and Meunier-Salaun, M.C. (1989) The influence of boar stimulation on puberty attainment in tethered and group penned gilts. *Theriogenology* 32, 949–957.

Prunier, A., Chopineau, M., Mountier, A.M. and Mormede, P. (1993) Patterns of plasma LH, FSH, oestradiol and corticosteroids from birth to the first oestrous cycle in Meishan gilts. *Journal of Reproduction and Fertility* 98, 313–319.

Rajamahendran, R., McElroy, L., Wallace, C., Ravindran, V. and Beames, R.M. (1992) Effects of boar exposure and/or progesterone priming on the induction and maintenance of cyclicity in prepubertal gilts receiving PMSG:HCG. *World Review of Animal Production* 27, 4, 8–9, 61–66.

Ramirez, D.V. and McCann, S.M. (1963) Comparison of the regulation of luteinizing hormone (LH) secretion in immature and adult rats. *Endocrinology* 72, 452–456.

Rampacek, G.B., Kraeling, R.R, and Ball, G.D. (1976) Luteal function in the hysterectomized prepubertal gilt. *Journal of Animal Science* 43, 792–794.

Rampacek, G.B., Kraeling, R.R. and Kiser, T.E. (1981) Delayed puberty in gilts in total confinement. *Theriogenology* 15, 491.

Robinson, J.J. (1990) Nutrition in the reproduction of farm animals. *Nutrition Research Reviews* 3, 253–276.

Scanlon, P.F. and Krishnamurthy, S. (1974) Puberty attainment in slaughterweight gilts in relation to month examined. *Journal of Animal Science* 39, 160.

Schilling, E. and Cerne, F. (1972) Induction and synchronization of oestrus in prepubertal gilts and anoestrous sows by a PMS/HCG compound. *Veterinary Record* 91, 471–474.

Schlegel, von W., Wahner, M. and Heinze, A. (1979) Studies into possible reduction of PMSG/HCG dose for cycle induction in the context of biotechnically induced puberty of gilts to use second oestrus. *Mh. Veterinary Medicine* 34, 187–188.

Schnurrbusch, U., Kauffold, J. and Richter, A. (1994) Synchronization of oestrus and ovulation in gilts by the prostaglandin $F_{2\alpha}$ analogue Tiaprost and gonadotrophins. *Tieraerztliche Umschau* 49, 238–248.

Schukken, Y.H., Buurman, J., Huirne, R.B.M., Willemse, A.H., Vernooy, J.C.M., Broek, J.Van Den and Verheuden, J.H.M. (1994) Evaluation of optimal age at first conception in gilts from data collected in commercial swine herds. *Journal of Animal Science* 72, 1387–1392.

Segal, D.H. and Baker, R.D. (1973) Maintenance of corpora lutea in prepubertal gilts. *Journal of Animal Science* 37, 62–67.

Signoret, J.P. (1989) Differentiation of sexual behaviour in young pigs. *21es Journees de la Recherche Porcine en France*, 129–132.

Sorensen, A.M.Jr, Thomas, W.W. and Gossett, J.W. (1961) A further study of the influence of level of energy intake and season on reproductive performance of gilts. *Journal of Animal Science* 20, 347–349.

Srikandakumar, A. and Downey, B.R. (1989) Induction of ovulation in gilts with cloprostenol. *Theriogenology* 32, 445–449.

Stephens, D.B. and Close, W.H. (1984) The influence of transportation, proximity of adults and other stimuli on the age of onset of first puberty in the gilt. *Cornell Vetarinarian* 74, 187–197.

Stephens, S., Duffy, P. and Boland, M.P. (1988) Induction of oestrus and AI in the gilt. *Proceedings of the 11th International Congress Animal Reproduction and AI* (Dublin), 4, Paper no. 461.

Stewart, A.H., Edwards, S.A. and Fowler, V.R. (1993) Synchronization of oestrus in once bred gilts using gonadotropins (PG600). *Proceedings of the British Society of Animal Production* (Winter Meeting), Paper no. 133.

Tess, M.W., Bennett, G.L. and Dickerson, G.E. (1983) Simulation of genetic changes in life cycle efficiency of pork production. I. A bioeconomic model. *Journal of Animal Science* 56, 336–353.

Van Lunen, T.A. and Aherne, F.X. (1987) Effect of long-term feed restriction on age at puberty of gilts. *Canadian Journal of Animal Science* 67, 797–801.

Walker, N. (1979) The occurrence of puberty in gilts reared in the presence or absence of growing boars. *Agricultural Record* 27, 1–4.

Walker, N., Kilpatrick, D.J. and Courtney, D.J. (1989) The effect of conception in gilts at puberty or second oestrus on reproductive performance over two parities. *Irish Journal of Agricultural Research* 28, 115–121.

Wise, M., Jones, A., Allrich, R. and Zimmernan, D.R. (1980) Influence of photoperiod and relocation-boar exposure stimuli on age at puberty in confinement-reared gilts. *Journal of Animal Science* 51 (Suppl. 1), 88.

Woods, C.F., Miller, G.F., McNew, R.W. and Rorie, R.W. (1995) Prepubertal luteinizing hormone levels as a predictor of onset of puberty in crossbred gilts. *Journal of Animal Science* 73 (Suppl. 1), 218.

Yang, H., Rodway, R.G. and Varley, M.A. (1987) The influence of different doses of oestradiol benzoate on the attainment of puberty in the gilt. *Animal Production* 44, 285–291.

Yang, H., Hughes, P.E., Rodway, R.G. and Varley, M.A. (1988) Effects of oestradiol benzoate injections, exposure to the boar and age on the attainment of puberty in the gilt. *Veterinary Record* 122, 257–259.

Zimmerman, D.R., Spies, H.G., Self, H.L. and Casida, L.E. (1960) Ovulation rate in swine as affected by increased energy intake just prior to ovulation. *Journal of Animal Science* 19, 295–301.

Zimmerman, D.R., Bourn, P. and Donovan, D. (1976) Effect of 'transport phenomenon' stimuli and boar exposure on puberty in gilts. *Journal of Animal Science* 42, 1362.

Index

Printed and bound by CPI Group (UK) Ltd, Croydon, CR0 4YY

16/09/2026

14963433-0001